T0381312

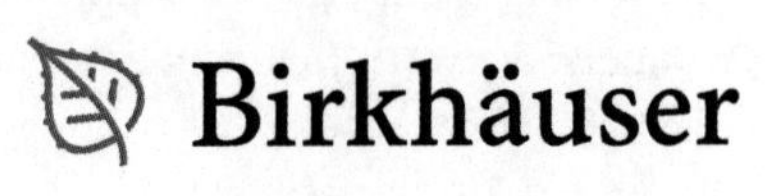
Birkhäuser

Frontiers in Mathematics

This series is designed to be a repository for up-to-date research results which have been prepared for a wider audience. Graduates and postgraduates as well as scientists will benefit from the latest developments at the research frontiers in mathematics and at the "frontiers" between mathematics and other fields like computer science, physics, biology, economics, finance, etc. All volumes are online available at SpringerLink.

Soon-Mo Jung

Aleksandrov-Rassias Problems on Distance Preserving Mappings

Soon-Mo Jung
Hallym University
Chuncheon, Korea (Republic of)

ISSN 1660-8046 ISSN 1660-8054 (electronic)
Frontiers in Mathematics
ISBN 978-3-031-77612-0 ISBN 978-3-031-77613-7 (eBook)
https://doi.org/10.1007/978-3-031-77613-7

This book is published under the imprint Birkhäuser, www.birkhauser-science.com by the registered company Springer Nature Switzerland AG
The registered company address is: Gewerbestrasse 11, 6330 Cham, Switzerland

Dedicated to Aleksandr D. Aleksandrov and
Themistocles M. Rassias

Preface

Isometry is not only widely used to describe the motion of rigid bodies that is a mixture of rotational and translational motions of solids, but is also an important tool in theoretical physics and geometry. The properties of isometry have been actively researched for a long time, and this theory is widely used in fields such as natural science, computer engineering, radiology, imaging science, and design.

It is generally recognized that the theory of isometries in Banach spaces originated from the paper [50] by S. Mazur and S. M. Ulam. Indeed, they proved the following theorem:

Every isometry from one real normed space onto another real normed space is affine.

This is the famous Mazur-Ulam theorem.

It is natural to ask whether the Mazur-Ulam theorem is true even without the "onto" assumption. In connection with this question, J. A. Baker [2] proved that the Mazur-Ulam theorem still holds if the "onto" condition of relevant isometry is removed and the condition on the range space is further strengthened:

If X is a real normed space and Y is a strictly convex real normed space, then every isometry $f : X \to Y$ is affine.

Many mathematicians have long been engaged in research to specify the conditions that determine isometry. The *Aleksandrov problem* falls within the scope of this research topic. In a paper [1], A. D. Aleksandrov posed the following problem:

Is any mapping that preserves a distance necessarily an isometry?

One of the goals in writing this book is to provide readers with a friendly explanation that draws their attention to the Aleksandrov problem.

It is quite interesting, however, that F. S. Beckman and D. A. Quarles [3] provided a partial solution to the Aleksandrov problem in 1953, when the Aleksandrov problem had not yet been raised. In fact, they proved the following theorem:

Let $\mathbb{E}^n$ denote the n-dimensional Euclidean space, where n is a fixed integer greater than 1. *If a mapping $f : \mathbb{E}^n \to \mathbb{E}^n$ preserves a distance $\rho > 0$, then f is an affine isometry.*

Some mathematicians have constructed several examples showing that there are mappings that preserve certain distances but are not isometries. These examples show that the Aleksandrov problem is not trivial, and we will present these examples in detail in this book.

If the range of the relevant mapping is a strictly convex real normed space, W. Benz and H. Berens [7] were able to generalize the Beckman-Quarles theorem as follows:

Let X be a real normed space with $\dim X > 1$ *and let Y be a real normed space which is strictly convex. Assume that N is a fixed integer greater than* 1. *If a distance $\rho > 0$ is contractive and $N\rho$ is extensive by a mapping $f : X \to Y$, then f is an affine isometry.*

In 1990, Th. M. Rassias [57] further generalized the idea of Benz and Berens by posing the following problem, which can certainly be considered an extension of the Aleksandrov problem:

Is a mapping that preserves two or more distances with non-integer ratios necessarily an isometry?

Such a problem is called the *Aleksandrov-Rassias problem*. A number of examples and counterexamples of the Aleksandrov-Rassias problem can be found in Sects. 4.2 and 4.3. This shows that the Aleksandrov-Rassias problem is not trivial. The most important purpose of this book is to increase readers' understanding by introducing the Aleksandrov-Rassias problem in detail.

This book is intended to give readers an overview of the process of solving the Aleksandrov-Rassias problem, which is still actively studied by many mathematicians, and to familiarize them with the details of the proof process. In addition, a lot of effort has been put into writing this book so that readers can easily understand the contents of the book and save readers the trouble of having to search the literature on their own.

This book consists of eight chapters, and we will now briefly describe the content of each chapter.

- Chapter 1 provides a brief introduction to the basic concepts and theorems of metric spaces, vector spaces, normed spaces, Banach spaces, inner product spaces, and Hilbert spaces, which are necessary to explain the subject of this book, the Aleksandrov-Rassias problems. In Sect. 1.6, we will systematically prove, from the point of view of a basic knowledge level, that ℓ^2 is a real Hilbert space. Readers who are familiar with the mathematical objects mentioned above can skip this chapter.

- In Chap. 2, we will examine in detail the Aleksandrov problem, one of the ways to find necessary and sufficient conditions for characterizing the isometries defined between Euclidean spaces. In the first section, we introduce the Mazur-Ulam theorem, which is believed to have triggered the study of the theory of isometries in Banach spaces. Section 2.2 presents Baker's theorem, which significantly generalizes the Mazur-Ulam theorem and improves its applicability. In Sect. 2.4, we devote considerable space to the detailed presentation of the proof of the Beckman-Quarles theorem, which directly motivated us to write this book.
- F. S. Beckman and D. A. Quarles solved the Aleksandrov problem for mappings from a Euclidean space into the same one. After weakening the result of the Beckman-Quarles theorem to make the proof easier, E. M. Schröder proved in 1979 that any mapping that preserves two distances ρ and 2ρ is an affine isometry. In a situation where the Beckman-Quarles theorem is known, Schröder's theorem by itself is of little significance. Schröder, however, presented a novel idea like the m-chain in the process of proving his theorem, and using this idea, W. Benz was able to significantly expand the Beckman-Quarles theorem. In Chap. 3, we present historically important theorems, the Schröder's theorem, the Benz's theorem, and the Benz-Berens theorem in detail.
- In the case where the domain and range spaces are the same Euclidean space whose dimension is greater than 1, the Aleksandrov problem has already been solved by the Beckman-Quarles theorem. W. Benz and H. Berens also solved the extended Aleksandrov problem under the additional conditions that the domain is a real normed space with dimension greater than 1, the range is a strictly convex real normed space, two distances are preserved, and that the ratio of the two distances is an integer. In Chap. 4, we investigate the Aleksandrov-Rassias problems, which focus on cases where the domain and range of the mapping involved differ, and cases where the ratio of two distances that are preserved is not an integer. By introducing interesting examples and counterexamples related to these topics in detail, we try to help readers easily grasp the core reality of the problem.
- In Chap. 5 and the next chapters, we will introduce in more detail the ideas and methods used to solve the Aleksandrov-Rassias problems for each case. Section 5.1 focuses on examining the Aleksandrov-Rassias problems, where the relevant mapping preserves two distances, 1 and $\sqrt{3}$. This is a special case among cases where the ratio of two distances is not an integer. In Sect. 5.2, we consider the Aleksandrov-Rassias problem where the relevant mapping preserves the distances 1 and $\sqrt{2}$. Section 5.3 friendly presents the conditions under which any mapping that preserves three distances is necessarily an isometry. As shown in Problem 4.2 (ii), this problem is also closely related to the Aleksandrov-Rassias problems.
- In Chap. 6, we will generalize the parallelogram law to all cases with a given number of points. In Sect. 6.1, we prove the short diagonals lemma which is a generalization of the parallelogram law for the given four points in the inner product space. Moreover, we

prove an inequality for the distances between any two points among the given six points. Section 6.2 is devoted to a generalization of the short diagonals lemma to an inequality for the distances among the given $2n$ points in the inner product space. In this section, we conceive and prove an inequality involving all distances between an even number of points. In Sect. 6.3, we use a new method other than the short diagonals lemma to prove an inequality for the distances between any two of the given five points. We introduce in Sect. 6.4 a general inequality that relates to the distances among the given n points.

- In Chap. 7, we will discuss ideas for partially solving the Aleksandrov-Rassias problems using the inequalities presented in Chap. 6. In the first section, we give a partial solution to the Aleksandrov-Rassias problems by using the inequality for the distances among six points presented in Sect. 6.1. Section 7.2 is devoted to proving that any mapping between real Hilbert spaces is an affine isometry if the distance 1 is preserved, $\frac{1}{\sqrt{2}}$ is contractive, and when $\sqrt{3}$ is extensive. In Sect. 7.3, we give a partial solution to the Aleksandrov-Rassias problems by proving that when the distance 1 is contractive and the golden ratio is extensive by a mapping defined between real Hilbert spaces, then this mapping is an affine isometry. In the last section of this chapter, we prove that a mapping between real Hilbert spaces is an affine isometry if the distances 1 and α are contractive, β is extensive, and if the distances 1, α, and β satisfy some suitable conditions.
- The Beckman-Quarles theorem states that every unit-distance preserving mapping $f : \mathbb{E}^n \to \mathbb{E}^n$ is an isometry if n is an integer greater than 1. Section 8.1 is devoted to the discussion of whether the Beckman-Quarles theorem also holds in rational n-spaces. It is known that every unit-distance preserving mapping $f : \mathbb{Q}^n \to \mathbb{Q}^n$ is an isometry if n is an even integer greater than 5 or an odd integer of the form $n = 2m^2 - 1$, where $m > 2$. Because it is beyond the scope of this book and due to space constraints, we must omit the interesting proofs of all theorems introduced in this section. In Sect. 8.2, we will discuss the theory of tensegrity structures that F. Rádo et al. used to partially solve the Aleksandrov-Rassias problems. Most of the content in this section comes from a paper by K. Bezdek and R. Connelly. Indeed, they were able to improve the result of Rádo et al. even further by refining the idea presented by Rádo et al. In Sects. 8.3 and 8.4, we provide some sufficient conditions for the Benz-Berens theorem and the Beckman-Quarles theorem to also hold in an open convex set. In the final section, we assume that the Beckman-Quarles theorem does not assume that the mapping preserves a certain distance, but rather a certain geometric figure. Over the past two decades, interesting results have been obtained on this topic, which will be systematically presented in the last section.

Aleksandrov-Rassias problems grow well in the fertile soil of real inner product spaces and have brought us plenty of fruit so far. With the hope that more mathematicians will turn their attention to this field of research and bring it to full fruition, this book is dedicated to the mathematicians who played a decisive role in pioneering the field: Aleksandr D. Aleksandrov and Themistocles M. Rassias. In addition, the author would like to dedicate this book to all mathematicians who are currently working or will work in this area in the future. The author would like to express his sincere gratitude to his wife, Min-Soon Lee, who devotedly supported him in writing this book for a long time despite her poor health.

Sejong, Republic of Korea
January, 2024
Soon-Mo Jung

Contents

Preliminaries

1

Abstract

In this chapter provides a brief introduction to the basic concepts and theorems of metric spaces, vector spaces, normed spaces, Banach spaces, inner product spaces, and Hilbert spaces, which are necessary to explain the subject of this book, the Aleksandrov–Rassias problems. For this purpose, we mainly refer to the book (Debnath and Mikusiński, *Introduction to Hilbert Spaces with Applications*, 2nd edn. Academic, New York, 2005) by L. Debnath and P. Mikusiński, among others. In Sect. 1.6, we will systematically prove, from the point of view of a basic knowledge level, that ℓ^2 is a real Hilbert space. For this end, we mainly refer to the book by R. H. Kasriel (*Undergraduate Topology*. W. B. Saunders, Philadelphia, 1971). Readers who are familiar with the mathematical objects mentioned above can skip this chapter.

1.1 Metric Spaces

A metric space is a set along with a metric for the set, and the metric is a mapping that generalizes the notion of a distance between any two elements of the set. In other literature, the elements of a set are often referred to as points, but in this book, we will refer to them interchangeably as points or elements.

Definition 1.1 Let X be a set, and let $d : X \times X \to [0, \infty)$ be a mapping that satisfies

(i) $d(x, y) = 0$ if and only if $x = y$;
(ii) $d(x, y) = d(y, x)$;
(iii) $d(x, y) \leq d(x, z) + d(z, y)$

S.-M. Jung, *Aleksandrov-Rassias Problems on Distance Preserving Mappings*, Frontiers in Mathematics, https://doi.org/10.1007/978-3-031-77613-7_1

for all $x, y, z \in X$. Then the mapping d is called a *metric* on X, and (X, d) is called a *metric space*. When it is clear what the metric d is, we often simply write X instead of (X, d). Let Y be a subset of X and let d^* be the restriction of the metric d to $Y \times Y$. Then (Y, d^*) is a metric space, and it is called a *subspace* of (X, d). In this case it is customary to write (Y, d) instead of (Y, d^*).

For each positive integer $n \in \mathbb{N}$, , let $\mathbb{R}^n$ be the n-dimensional space of all ordered n-tuples of real numbers. We define the mapping $d_e : \mathbb{R}^n \times \mathbb{R}^n \to [0, \infty)$ by

$$d_e(x, y) = \left(\sum_{i=1}^{n} (x_i - y_i)^2 \right)^{1/2}$$

for all $x = (x_1, x_2, \ldots, x_n), y = (y_1, y_2, \ldots, y_n) \in \mathbb{R}^n$. Then, d_e satisfies all the conditions in Definition 1.1, i.e., it is a metric on $\mathbb{R}^n$, which is called the *Euclidean metric* on $\mathbb{R}^n$.

Any set X can be viewed as a metric space in a natural way. For example, we define the mapping $d : X \times X \to [0, \infty)$ by

$$d(x, y) = \begin{cases} 0 \text{ (for } x = y), \\ 1 \text{ (for } x \neq y) \end{cases}$$

for all $x, y \in X$. It is easy to check that d is a metric on the set X. This metric is called the *discrete metric*.

Suppose x is an element of a metric space (X, d) and $r > 0$ is a real number. Define an *open ball* and a *closed ball* with center at x and radius r as

$$B_r(x) = \{y \in X : d(y, x) < r\} \text{ and } \overline{B}_r(x) = \{y \in X : d(y, x) \leq r\},$$

respectively.

Definition 1.2 Assume that K and U are subsets of a metric space (X, d).

(i) U is said to be *open* if and only if for every $x \in U$, there exists a real number $\varepsilon > 0$ such that $B_\varepsilon(x) \subset U$.

(ii) K is said to be *closed* if and only if its complement $X \setminus K$ is open.

We will often write "U is open in X" and "K is closed in X" instead of "U is an open subset of X" and "K is a closed subset of X," respectively.

Lemma 1.3 *Assume that x is an element of a metric space (X, d) and r is a positive real number.*

(*i*) *The open ball* $B_r(x)$ *is open in* X.
(*ii*) *The closed ball* $\overline{B}_r(x)$ *is closed in* X.

Proof

(*i*) For any element $y \in B_r(x)$, we see that $d(y, x) < r$ and we set $\varepsilon = r - d(y, x) > 0$. If z is an arbitrary element of $B_\varepsilon(y)$, then we have

$$d(z, x) \leq d(z, y) + d(y, x) < \varepsilon + d(y, x) = r - d(y, x) + d(y, x) = r,$$

which implies that $z \in B_r(x)$, i.e., $y \in B_\varepsilon(y) \subset B_r(x)$. Thus, according to Definition 1.2 (*i*), $B_r(x)$ is an open subset of X.
(*ii*) We encourage the reader to do his/her own proof of (*ii*). □

Definition 1.4 Let $\mathcal{T}$ be a collection of subsets of a set X. The collection $\mathcal{T}$ is called a *topology* for X if and only if $\mathcal{T}$ satisfies the following conditions:

(*i*) $\emptyset \in \mathcal{T}$ and $X \in \mathcal{T}$;
(*ii*) If $U_1 \in \mathcal{T}$ and $U_2 \in \mathcal{T}$, then $U_1 \cap U_2 \in \mathcal{T}$;
(*iii*) If $\mathcal{U}$ is an arbitrary sub-collection of $\mathcal{T}$, then $\bigcup_{U \in \mathcal{U}} U \in \mathcal{T}$.

In this case $(X, \mathcal{T})$ is called a *topological space*. When it is clear what the topology $\mathcal{T}$ is, we often simply write X instead of $(X, \mathcal{T})$. Let Y be a subset of X and let $\mathcal{T}_Y$ be the collection

$$\mathcal{T}_Y = \{U \cap Y : U \in \mathcal{T}\}.$$

Then $\mathcal{T}_Y$ is called the *relative topology* for Y induced by $\mathcal{T}$. Moreover, $(Y, \mathcal{T}_Y)$ is called a *subspace* of $(X, \mathcal{T})$. In particular, a subset U of X is said to be *open* in X if and only if $U \in \mathcal{T}$, and a subset K of X is said to be *closed* in X if and only if $X \setminus K \in \mathcal{T}$.

We denote by $\mathcal{P}(X)$ the *power set* of a set X, i.e., the collection of all subsets of X. Then $\mathcal{P}(X)$ is a topology for X, which is called the *discrete topology* for X. Let $\mathcal{T} = \{\emptyset, X\}$. Then $\mathcal{T}$ is also a topology for X, which is called the *trivial topology* or the *indiscrete topology* for X. In general, any set can have at least two topologies, the discrete topology and the indiscrete topology.

Let (X, d) be a metric space, and let $\mathcal{T}(d)$ be the collection of all open subsets of X defined by Definition 1.2. It is then easy to prove that $(X, \mathcal{T}(d))$ is a topological space.

Definition 1.5 Let (X, d) be a metric space and $\mathcal{T}(d)$ the collection of all open subsets U of X such that each element x of U satisfies the condition, that $B_\varepsilon(x) \subset U$ for some

$\varepsilon > 0$. Then the collection $\mathcal{T}(d)$ is a topology for X, and it is called the *topology for X generated by d*.

A topological space $(X, \mathcal{T})$ is said to be *metrizable* if and only if there exists a metric d on X whose collection $\mathcal{T}(d)$ is exactly the topology $\mathcal{T}$. For this reason, we often treat the metric space (X, d) as a topological space $(X, \mathcal{T}(d))$. We note that the discrete topology is generated by the discrete metric, but the indiscrete topology cannot be generated by any metric.

Definition 1.6 Let K and U be subsets of a topological space X.

(i) A point $x \in X$ is called an *interior point* of U if and only if there exists an open subset V of X such that $x \in V \subset U$.
(ii) A point $x \in X$ is called a *limit point* of K if and only if every open set containing x intersects K in a point distinct from x.
(iii) The set of all interior points of U is called the *interior* of U and is denoted by U°.
(iv) The union of K and the set of all its limit points is called the *closure* of K and is denoted by $\overline{K}$.

Theorem 1.7 *Assume that S is a subset of a topological space X.*

(i) *The interior of S is the union of all open subsets of X that are included in S, i.e.,*

$$S^\circ = \bigcup \{U : U \text{ is an open subset of } X \text{ included in } S\}.$$

(ii) *The closure of S is the intersection of all closed subsets of X that include S, i.e.,*

$$\overline{S} = \bigcap \{K : K \text{ is a closed subset of } X \text{ including } S\}.$$

Let $\mathbb{N}$ be the set of all positive integers, and let $\mathbb{N}_0 = \mathbb{N} \cup \{0\}$. A sequence $\{x_i\}_{i \in \mathbb{N}}$ of elements of a subset S of a topological space X is said to *converge* to an element x of X if and only if for every open subset U of X containing x, there exists a positive integer N such that $x_i \in U \cap S$ for all integers $i \geq N$.

The proof of the following theorem is left as an exercise for the reader.

Theorem 1.8 *The closure of a subset S of a topological space is the set of limits of all convergent sequences of elements of S.*

Definition 1.9 Let X be a metric space.

(i) A subset D of X is called a *dense* subset of X (or *dense* in X) if and only if $\overline{D} = X$.

(ii) X is called *separable* if and only if there exists a countable subset of X that is dense in X.

The following characterization of denseness is easy to prove, and its proof will be left as an exercise.

Theorem 1.10 *Assume that D is a subset of a metric space X. Then D is dense in X if and only if each nonempty open subset of X intersects D.*

In topology, a homeomorphism is a bijective and continuous mapping between topological spaces that has a continuous inverse mapping. Homeomorphisms preserve all the topological properties of a given space. Therefore, two topological spaces with a homeomorphism between them are called homeomorphic and are equal from a topological point of view.

Definition 1.11 Let X and Y be topological spaces. A mapping $f : X \to Y$ is called a *homeomorphism* (or a *topological mapping*) if and only if f satisfies the following conditions:

(i) f is bijective;
(ii) f is continuous;
(iii) f^{-1} is continuous.

Two topological spaces X and Y are called *homeomorphic* (or *topologically equivalent*) if and only if there is a homeomorphism from X onto Y.

Let $\mathcal{T}_d$ be the discrete topology, and let $\mathcal{T}_e$ be the *usual topology* for $\mathbb{R}$, i.e., $\mathcal{T}_e = \mathcal{T}(d_e)$, where $d_e(x, y) = |x - y|$ for all $x, y \in \mathbb{R}$. Then the identity mapping $\mathrm{id} : (\mathbb{R}, \mathcal{T}_d) \to (\mathbb{R}, \mathcal{T}_e)$ is continuous and bijective. However, it is not a homeomorphism because $\mathrm{id}^{-1} : (\mathbb{R}, \mathcal{T}_e) \to (\mathbb{R}, \mathcal{T}_d)$ is not continuous.

Example 1.12 Let $\mathcal{T}$ denote the usual topology for $\mathbb{R}$. We assert that $(\mathbb{R}, \mathcal{T})$ is homeomorphic to $((-1, 1), \mathcal{T}_{(-1,1)})$: We define a mapping $f : \mathbb{R} \to (-1, 1)$ by

$$f(x) = \frac{x}{1 + |x|}$$

for all $x \in \mathbb{R}$. Then f is continuous and bijective. Moreover, the inverse mapping $f^{-1}(y) = \frac{y}{1-|y|}$ is continuous. Thus, f is a homeomorphism.

Suppose that if a topological space X has a property, then every topological space homeomorphic to X also has this property. Then we say that this property is a topological property. Informally, a topological property is a property of space that can be expressed in terms of open sets.

Definition 1.13 A property of a topological space is called a *topological property* (or a *topological invariant*) if and only if it is invariant under homeomorphisms.

According to Example 1.12, the property that "the topological space has a finite diameter" is not a topological property since the diameter of $\mathbb{R}$ is infinite, but that of $(-1, 1)$ is finite.

Intuitively, we would like to say that a set is connected when it "hangs" as a whole. Whatever definition we choose for connected space, we would certainly require, for example, that the real line $\mathbb{R}$ has this property. Note that there are many ways to decompose the real line into a pair of nonempty disjoint subsets. However, it can be shown that $\mathbb{R}$ is not the union of two nonempty disjoint open sets.

Definition 1.14 Let $(X, \mathcal{T})$ be a topological space. Then $(X, \mathcal{T})$ is *connected* if and only if X is not the union of two nonempty disjoint open subsets. A subset S of X is called connected if and only if $(S, \mathcal{T}_S)$ is a connected space. If a set is not connected, we say that it is *disconnected.*

Obviously, the empty set $\emptyset$ is connected. Moreover, all single-element sets are connected, but two-element sets may not be. For example, let $\mathcal{T}_d$ be the discrete topology for the set $X = \{0, 1\}$. Then the topological space $(X, \mathcal{T}_d)$ is not connected because $X = \{0\} \cup \{1\}$ with $\{0\}, \{1\} \in \mathcal{T}_d$.

Theorem 1.15 *A topological space X is connected if and only if there is no subset S of X with $\emptyset \neq S \neq X$ that is both open and closed.*

Proof We will prove that a topological space $(X, \mathcal{T})$ is disconnected if and only if there is a subset S of X with $\emptyset \neq S \neq X$ that is both open and closed.

First, assume that $(X, \mathcal{T})$ is disconnected. There are two nonempty disjoint open subsets U_1 and U_2 such that $X = U_1 \cup U_2$. If we set $S = U_1$, then $\emptyset \neq S \neq X$, $S \in \mathcal{T}$, and $S = X \setminus U_2$ is closed. That is, there exists a subset S of X with $\emptyset \neq S \neq X$ that is both open and closed.

Now we assume that there exists a subset S of X with $\emptyset \neq S \neq X$ that is both open and closed. If we set $U_1 = S$ and $U_2 = X \setminus S$, then U_1 and U_2 are nonempty open subsets of X, U_1 and U_2 are disjoint, and $X = U_1 \cup U_2$. That is, $(X, \mathcal{T})$ is disconnected. □

Let $X = (0, 1) \cup (2, 3)$ be a topological subspace of $(\mathbb{R}, \mathcal{T}_e)$, where $\mathcal{T}_e$ is the usual topology. Then $(0, 1)$ is open and closed in X, since $(0, 1) = (0, 1) \cap X = [0, 1] \cap X$. In addition, $\emptyset \neq (0, 1) \neq X$. Hence, X is disconnected.

We note that a mapping $f : X \to Y$ is continuous if and only if $f^{-1}(K)$ is closed in X for each set K that is closed in Y.

Theorem 1.16 *Let $(X, \mathcal{T})$ be a topological space and let $Z = \{0, 1\}$ have the discrete topology $\mathcal{T}_d$. Then X is connected if and only if every continuous mapping $f : X \to Z$ is constant.*

Proof First, we assume that the topological space $(X, \mathcal{T})$ is connected. Suppose $f : X \to Z$ is a continuous mapping and $z \in f(X)$. Since f is continuous and $\{z\} \in \mathcal{T}_d$, it holds that $f^{-1}(\{z\}) \neq \emptyset$, $f^{-1}(\{z\}) \in \mathcal{T}$, and $f^{-1}(\{z\})$ is closed. It then follows from Theorem 1.15 that $f^{-1}(\{z\}) = X$, i.e., $f(X) = \{z\}$.

Now, we assume that every continuous mapping $f : X \to Z$ is constant. On the contrary, assume that X is disconnected. On account of Theorem 1.15, there is a subset S of X with $\emptyset \neq S \neq X$ that is both open and closed. We consider the characteristic mapping $\chi_S : X \to Z$ defined by

$$\chi_S(x) = \begin{cases} 1 \text{ (for } x \in S), \\ 0 \text{ (for } x \notin S). \end{cases}$$

Since $\chi_S^{-1}(\emptyset) = \emptyset$, $\chi_S^{-1}(\{0\}) = X \setminus S$, $\chi_S^{-1}(\{1\}) = S$, and since $\chi_S^{-1}(Z) = X$, χ_S is a continuous mapping. However, χ_S is not constant, which contradicts our assumption. Therefore, X has to be connected. □

Remark 1.17 A subset S of $\mathbb{R}$ is connected if and only if S is an interval.

Proof Assume that S is connected and non-degenerated (i.e., S has two or more elements). On the contrary, suppose S is not an interval. We can then choose a point $x \notin S$ such that $(-\infty, x) \cap S \neq \emptyset$ and $(x, \infty) \cap S \neq \emptyset$. Then, we have

$$S = \big((-\infty, x) \cap S\big) \cup \big((x, \infty) \cap S\big).$$

Let $Z = \{0, 1\}$ have the discrete topology $\mathcal{T}_d$. We define a mapping $f : S \to Z$ by

$$f(s) = \begin{cases} 0 \text{ (for } s \in (-\infty, x) \cap S), \\ 1 \text{ (for } s \in (x, \infty) \cap S). \end{cases}$$

It is easy to check that f is continuous, and it is not a constant mapping. It then follows from Theorem 1.16 that S is disconnected, a contradiction. Therefore, we conclude that if S is a connected subset of $\mathbb{R}$, then it is an interval. We note that every single-element set is connected and a degenerated interval.

Assume that S is a non-degenerated interval and $f : S \to Z$ is a continuous mapping. Then, $f : S \to \mathbb{R}$ is also continuous and $f(S) \subset \{0, 1\}$. Hence, f has to be constant. Therefore, it follows from Theorem 1.16 that S is connected. □

Theorem 1.18 *Assume that X and Y are topological spaces and X is connected. If there is a surjective continuous mapping $f : X \to Y$, then Y is connected.*

Proof Assume that X is connected, $Z = \{0, 1\}$ has the discrete topology, and $g : Y \to Z$ is an arbitrary continuous mapping. Then, $g \circ f : X \to Z$ is a continuous mapping. According to Theorem 1.16, the composition $g \circ f$ is a constant mapping.

We claim that $g : Y \to Z$ is a constant mapping. Assume that y_1 and y_2 are arbitrary elements of Y. Since f is surjective, there are $x_1, x_2 \in X$ such that $y_1 = f(x_1)$ and $y_2 = f(x_2)$. Since $g \circ f$ is a constant mapping, it follows that $g(y_1) = (g \circ f)(x_1) = (g \circ f)(x_2) = g(y_2)$, i.e., g is a constant mapping.

Finally, it follows from Theorem 1.16 that Y is connected. □

Theorem 1.18 reminds us that connectedness is a topological property.

Remark 1.19 The connectedness is a topological property. In view of Remark 1.17, every line ℓ in $\mathbb{R}^n$ is connected since ℓ is homeomorphic to $\mathbb{R}$.

Theorem 1.20 *Suppose X is a topological space and Γ is a class of connected subsets of X such that*

(i) $A \cap B \neq \emptyset$ *for all* $A, B \in \Gamma$;
(ii) *there exists a* $C \in \Gamma$ *such that* $A \cap C \neq \emptyset$ *for all* $A \in \Gamma$.

Then $\bigcup_{S \in \Gamma} S$ *is connected.*

Proof We set $U = \bigcup_{S \in \Gamma} S$ and assume that $Z = \{0, 1\}$ has the discrete topology and $f : U \to Z$ is an arbitrary continuous mapping.

We assert that f is constant. For every $S \in \Gamma$, the restriction $f|_S : S \to Z$ of f to S is continuous and S is connected. Then, by Theorem 1.16, $f|_S$ is constant for every $S \in \Gamma$. Using (i) and (ii), we conclude that f is constant. Therefore, using Theorem 1.16 again, we conclude that U is connected. □

Remark 1.21 $\mathbb{R}^n$ is connected.

Proof Let Γ be the class of all lines ℓ through the origin. Due to Remark 1.19, each line ℓ through the origin is connected. Since $\mathbb{R}^n = \bigcup_{\ell \in \Gamma} \ell$, it follows from Theorem 1.20 that $\mathbb{R}^n$ is connected. □

Determining whether two topological spaces are homeomorphic is one of the important problems in topology. To prove that two spaces are not homeomorphic, it suffices to find a topological property that they do not have in common.

Remark 1.22 For any integer $n > 1$, $\mathbb{R}^n$ is not homeomorphic to $\mathbb{R}$.

Proof Assume that there is a homeomorphism $f : \mathbb{R} \to \mathbb{R}^n$. Then, $f|_{\mathbb{R}\setminus\{0\}} : \mathbb{R} \setminus \{0\} \to \mathbb{R}^n \setminus \{f(0)\}$ is also a homeomorphism. However, it contradicts the fact that connectedness is a topological property. In fact, $\mathbb{R} \setminus \{0\}$ is disconnected but $\mathbb{R}^n \setminus \{f(0)\}$ is connected. □

Assume that $f : X \to Y$ is a mapping between topological spaces X and Y. We note that f is continuous if and only if $f(\overline{S}) \subset \overline{f(S)}$ for any subset S of X.

Theorem 1.23 *If S is a connected subset of a topological space X, then so is $\overline{S}$.*

Proof Assume that $Z = \{0, 1\}$ has the discrete topology, $S \neq \emptyset$, and $f : \overline{S} \to Z$ is an arbitrary continuous mapping. Then, $f|_S : S \to Z$ is also a continuous mapping. Moreover, since Z has the discrete topology, $f(S)$ is closed in Z, i.e., $f(S) = \overline{f(S)} \supset f(\overline{S})$. By Theorem 1.16, $f(S)$ contains exactly one point. Hence, $f(\overline{S})$ contains exactly one point, which implies that f is constant. Finally, in view of Theorem 1.16, we conclude that $\overline{S}$ is connected. □

The property of the connectedness of intervals allows us to apply the intermediate value theorem to all real-valued continuous mappings defined in an interval.

Theorem 1.24 *Let X be a topological space. Then X is connected if and only if the intermediate value theorem holds for every continuous mapping $f : X \to \mathbb{R}$.*

Proof Assume that X is connected. According to Theorem 1.18, $f(X)$ is a connected subset of $\mathbb{R}$. Thus, it follows from Remark 1.17 that $f(X)$ is an interval. Hence, for all $x_1, x_2 \in X$ and $c \in \mathbb{R}$ with $f(x_1) \leq c \leq f(x_2)$, there exists an $x_0 \in X$ such that $c = f(x_0)$.

Conversely, we assume that the intermediate value theorem holds for every continuous mapping $f : X \to \mathbb{R}$. On the contrary, suppose X is disconnected. Then there are nonempty disjoint open subsets U_1 and U_2 of X such that $X = U_1 \cup U_2$. We define the mapping $f : X \to \mathbb{R}$ by $f(U_1) = \{0\}$ and $f(U_2) = \{1\}$. Then, f is a continuous mapping. The intermediate value theorem does not hold for this continuous mapping $f : X \to \mathbb{R}$, which contradicts our assumption. Therefore, X has to be connected. □

1.2 Vector Spaces

Let $\mathbb{K}$ be either $\mathbb{R}$ or $\mathbb{C}$, where $\mathbb{R}$ stands for the set of all real numbers and $\mathbb{C}$ stands for the set of all complex numbers. Throughout this book, we denote by $\mathbb{Q}$ the set of all rational numbers, by $\mathbb{Z}$ the set of all integers, by $\mathbb{N}$ the set of all positive integers, and by $\mathbb{N}_0$ the set of all nonnegative integers.

Definition 1.25 A nonempty set V is called a *vector space* (or a *linear space*) over $\mathbb{K}$ if there are two operations, called *vector addition* and *scalar multiplication*, such that the following conditions are satisfied:

(i) $x + y = y + x$ for all $x, y \in V$;
(ii) $(x + y) + z = x + (y + z)$ for all $x, y, z \in V$;
(iii) For any $x, y \in V$, there exists a $z \in V$ such that $x + z = y$;
(iv) $\alpha(\beta x) = (\alpha\beta)x$ for all $\alpha, \beta \in \mathbb{K}$ and any $x \in V$;
(v) $(\alpha + \beta)x = \alpha x + \beta x$ for all $\alpha, \beta \in \mathbb{K}$ and any $x \in V$;
(vi) $\alpha(x + y) = \alpha x + \alpha y$ for all $\alpha \in \mathbb{K}$ and all $x, y \in V$;
(vii) $1x = x$ for all $x \in V$.

Every element of $\mathbb{K}$ is called a *scalar* and each element of V is called a *vector*. If $\mathbb{K} = \mathbb{R}$, then V is called a *real vector space*, and if $\mathbb{K} = \mathbb{C}$, then V is called a *complex vector space*.

Definition 1.26 A subset W of a vector space V over $\mathbb{K}$ is called a *vector subspace* or a *subspace* if and only if $\alpha x + \beta y \in W$ for all $\alpha, \beta \in \mathbb{K}$ and $x, y \in W$. That is, the sum of two elements of W and the product of an element of W by a scalar belong to W. A subset W of a vector space V is called a *proper subspace* of V if W is a subspace of V and $W \neq V$.

We note that any subspace W of a vector space V is a vector space itself, which implies that every *linear combination* of elements of W again belongs to W.

Definition 1.27 Let V be a vector space.

(i) A finite collection $\{x_1, x_2, \ldots, x_n\}$ of elements of V is said to be *linearly independent* if and only if $\alpha_1 = \alpha_2 = \cdots = \alpha_n = 0$ is the unique solution of the linear equation $\alpha_1 x_1 + \alpha_2 x_2 + \cdots + \alpha_n x_n = 0$.
(ii) An infinite collection $\mathcal{A}$ of elements of V is called *linearly independent* if and only if every finite sub-collection of $\mathcal{A}$ is linearly independent.
(iii) A collection of elements of V is called *linearly dependent* if and only if it is not linearly independent.

A collection $\mathcal{A}$ of elements of a vector space V is linearly independent if and only if no member x of the collection is a linear combination of a finite number of members of $\mathcal{A}$ different from x.

Definition 1.28 Let A be a subset of a vector space V over $\mathbb{K}$. We denote by span$A$ the set of all finite linear combinations of members of A, i.e.,

$$\operatorname{span}A = \big\{\alpha_1 x_1 + \alpha_2 x_2 + \cdots + \alpha_n x_n : n \in \mathbb{N},\ \alpha_i \in \mathbb{K},\ x_i \in A$$
$$\text{for all } i \in \{1, 2, \ldots, n\}\big\}.$$

Then span$A$ is a vector subspace of V and it is called the *space spanned by* A.

It is easy to see that the space span$A$ (spanned by A) is the smallest vector subspace of V that contains A.

Definition 1.29 Let B be a subset of a vector space V. Then B is called a *basis* of V if and only if B is linearly independent and $\operatorname{span}B = V$.

In general, a vector space V has multiple bases, but the number of vectors in each basis is the same. In other words, if a basis of V has exactly n vectors, then every other basis also has exactly n vectors. In this case, n is called the *dimension* of V, and we write $\dim V = n$.

1.3 Normed Spaces

The notion of norm in the vector space is an abstract generalization of length in the Euclidean space. In fact, we define the norm axiomatically as a real-valued mapping that satisfies certain conditions.

Definition 1.30 Let V be a vector space over $\mathbb{K}$. A mapping $\|\cdot\| : V \to \mathbb{R}$ is said to be a *norm* if and only if it possesses the following three properties:

(i) $\|x\| = 0$ if and only if $x = 0$;
(ii) $\|\alpha x\| = |\alpha|\|x\|$ for all $x \in V$ and $\alpha \in \mathbb{K}$;
(iii) $\|x+y\| \le \|x\|+\|y\|$ for all $x, y \in V$. This inequality is called the *triangle inequality*.

Since $\|x\| = \frac{1}{2}(\|x\| + \|-x\|) \ge \frac{1}{2}\|x + (-x)\| = \frac{1}{2}\|0\| = 0$, it holds that $\|x\| \ge 0$ for any $x \in V$. The *inverse triangle inequality*

$$\|x - y\| \ge \big|\|x\| - \|y\|\big|,$$

for all $x, y \in V$, is an elementary consequence of the triangle inequality that gives lower bounds instead of upper bounds.

The mapping $\|\cdot\| : \mathbb{R}^n \to \mathbb{R}$ defined by $\|x\| = \sqrt{x_1^2 + x_2^2 + \cdots + x_n^2}$, for all $x = (x_1, x_2, \ldots, x_n) \in \mathbb{R}^n$, is a norm on $\mathbb{R}^n$. This norm is called the *Euclidean norm*.

Definition 1.31 A vector space with a norm is called a *normed space*.

Several norms can be defined on one vector space. For example, if we define $\|x\|_1 = |x_1| + |x_2| + \cdots + |x_n|$ for all $x = (x_1, x_2, \ldots, x_n) \in \mathbb{R}^n$, then this mapping $\|\cdot\|_1 : \mathbb{R}^n \to \mathbb{R}$ is a norm on the vector space $\mathbb{R}^n$. Moreover, $\|x\|_2 = \sqrt{|x_1|^2 + |x_2|^2 + \cdots + |x_n|^2}$, for $x = (x_1, x_2, \ldots, x_n) \in \mathbb{R}^n$, is the Euclidean norm. Similarly, the mapping $\|x\|_\infty = \max\{|x_1|, |x_2|, \ldots, |x_n|\}$ is also a norm on $\mathbb{R}^n$. This norm $\|\cdot\|_\infty$ is called the *sup-norm*. Therefore, to define a normed space, we have to specify both the vector space and the norm. We denote a normed space as $(V, \|\cdot\|)$, where V is a vector space and $\|\cdot\|$ is a norm defined on V. However, if the norm given in the normed space $(V, \|\cdot\|)$ is clear, then the normed space can be simply written as V.

Throughout this book, we will use the notation $\{x_i\}$ or $\{x_1, x_2, \ldots\}$ to denote the sequence whose ith term is x_i.

Definition 1.32 Let $(V, \|\cdot\|)$ be a normed space. A sequence $\{x_i\}$ of vectors in V is called a *Cauchy sequence* if and only if for each $\varepsilon > 0$ there exists a positive real number M_ε such that $\|x_i - x_j\| < \varepsilon$ for all $i, j > M_\varepsilon$.

If a sequence $\{x_i\}$ of vectors in a normed space $(V, \|\cdot\|)$ converges to a vector $x \in V$, i.e., $\|x_i - x\| \to 0$ as $i \to \infty$, then

$$\|x_i - x_j\| \leq \|x_i - x\| + \|x_j - x\| \to 0$$

as $i, j \to \infty$. Hence, every convergent sequence of vectors in a normed space is a Cauchy sequence. However, the reverse is not usually true.

Lemma 1.33 *If $\{x_i\}$ is a Cauchy sequence in a normed space $(V, \|\cdot\|)$, then the sequence $\{\|x_i\|\}$ converges.*

Proof Since $|\|x_i\| - \|x_j\|| \leq \|x_i - x_j\|$ for all $i, j \in \mathbb{N}$, the sequence $\{\|x_i\|\}$ is a Cauchy sequence in $\mathbb{R}$. Thus, we can conclude that $\{\|x_i\|\}$ converges. □

According to Lemma 1.33, every Cauchy sequence in a normed space is bounded.

Definition 1.34 Let $(V, \|\cdot\|)$ be a normed space.

(i) V is called *complete* if and only if every Cauchy sequence in V converges to an element of V.
(ii) Each complete normed space is called a *Banach space*.

The n-dimensional Euclidean space $\mathbb{E}^n$ is an example of the Banach space.

Definition 1.35 Let $(V, \|\cdot\|)$ be a normed space.

(i) A series $\sum_{i=1}^{\infty} x_i$ *converges* in V if and only if there exists an element x of V such that

$$\left\| \sum_{i=1}^{n} x_i - x \right\| \to 0 \quad \text{as} \quad n \to \infty.$$

In this case, we write $\sum_{i=1}^{\infty} x_i = x$.

(ii) When $\sum_{i=1}^{\infty} \|x_i\| < \infty$, the series $\sum_{i=1}^{\infty} x_i$ is called *absolutely convergent*.

It is surprising that even if a series converges absolutely, it might not converge. The following theorem shows that completeness in a normed space is equivalent to the statement that every absolutely convergent series converges.

Theorem 1.36 *A normed space $(V, \|\cdot\|)$ is complete if and only if every absolutely convergent series in V converges in V.*

Proof Assume that $(V, \|\cdot\|)$ is a complete normed space and $x_i \in V$ for all $i \in \mathbb{N}$ such that $\sum_{i=1}^{\infty} \|x_i\| < \infty$, i.e., the series $\sum_{i=1}^{\infty} x_i$ converges absolutely. We now define

$$s_n = x_1 + x_2 + \cdots + x_n$$

for all $n \in \mathbb{N}$.

We will prove that $\{s_n\}$ is a Cauchy sequence in V. Assume that ε is an arbitrary positive real number and N is a positive integer that satisfy

$$\sum_{n=N+1}^{\infty} \|x_n\| < \varepsilon.$$

Then, using the triangle inequality, we obtain

$$\|s_m - s_n\| = \|x_{m+1} + x_{m+2} + \cdots + x_n\| \le \sum_{k=m+1}^{\infty} \|x_k\| < \varepsilon$$

for all $m, n \in \mathbb{N}$ with $n > m > N$, which implies that $\{s_n\}$ is a Cauchy sequence in V. Since V is complete, the Cauchy sequence $\{s_n\}$ converges in V, which implies that the series $\sum_{i=1}^{\infty} x_i$ converges in V.

Conversely, we assume that every absolutely convergent series converges and $\{x_i\}$ is an arbitrary Cauchy sequence in V. Then, for any positive integer k, there exists a positive integer p_k such that $\|x_i - x_j\| < \frac{1}{2^k}$ for all integers $i, j > p_k$. We may assume that the sequence $\{p_k\}$ is strictly increasing.

Since the series $\sum_{k=1}^{\infty} (x_{p_{k+1}} - x_{p_k})$ is absolutely convergent, it converges by our assumption. Since

$$x_{p_k} = x_{p_1} + (x_{p_2} - x_{p_1}) + \cdots + (x_{p_k} - x_{p_{k-1}}) = x_{p_1} + \sum_{i=1}^{k-1} (x_{p_{i+1}} - x_{p_i}),$$

the sequence $\{x_{p_k}\}$ converges to an element x of V. Thus, we obtain

$$\|x_i - x\| \le \|x_i - x_{p_i}\| + \|x_{p_i} - x\| \to 0$$

as $i \to \infty$, i.e., the Cauchy sequence $\{x_i\}$ converges in V, which implies that $(V, \|\cdot\|)$ is complete. □

We note that every normed space $(V, \|\cdot\|)$ is a metric space (V, d), where the metric $d : V \times V \to \mathbb{R}$ is defined as $d(x, y) = \|x - y\|$ for all $x, y \in V$. In this case, the metric d is said to be the *metric generated by the norm* $\|\cdot\|$.

We now introduce a well-known theorem in topology.

Lemma 1.37 *Every closed subset of a complete normed space is complete.*

Proof Let $(V, \|\cdot\|)$ be a complete normed space. If we define a mapping $d : V \times V \to \mathbb{R}$ by $d(x, y) = \|x - y\|$, then (V, d) is a metric space. Thus, (V, d) is a complete metric space. It is well known that a closed subset W of a complete metric space V is complete. □

Using this lemma, we can easily prove the following theorem.

Theorem 1.38 *A closed subspace of a Banach space is a Banach space.*

Proof Let W be a closed subspace of a Banach space V. Then W is a normed space, which is closed in V. According to Lemma 1.37, W is complete as a closed subset of a complete normed space V. Therefore, W is a Banach space. □

1.4 Inner Product Spaces

An inner product space is a vector space over $\mathbb{K}$ with an operation called an inner product. The inner product of two vectors x and y in the space is a scalar, denoted with angle brackets such as $\langle x, y\rangle$.

We denote by $\overline{\alpha}$ the *complex conjugate* of the scalar α. We see that $\overline{\alpha} = \alpha$ for every $\alpha \in \mathbb{R}$.

Definition 1.39 Let V be a vector space over $\mathbb{K}$. A mapping $\langle\cdot,\cdot\rangle : V \times V \to \mathbb{K}$ is said to be an *inner product* for V if and only if the mapping has the the following four properties:

(i) $\langle x, y\rangle = \overline{\langle y, x\rangle}$;
(ii) $\langle \alpha x + \beta y, z\rangle = \alpha\langle x, z\rangle + \beta\langle y, z\rangle$;
(iii) $\langle x, x\rangle \geq 0$;
(iv) $\langle x, x\rangle = 0$ if and only if $x = 0$

for all $x, y, z \in V$ and $\alpha, \beta \in \mathbb{K}$. A vector space with an inner product is called an *inner product space* or a *pre-Hilbert space*.

It follows from Definition 1.39 (i) that $\langle x, x\rangle = \overline{\langle x, x\rangle}$, which implies that $\langle x, x\rangle$ is a real number for any $x \in V$. According to Definition 1.39 (i) and (ii), we have

$$\langle x, \alpha y + \beta z\rangle = \overline{\langle \alpha y + \beta z, x\rangle} = \overline{\alpha}\langle x, y\rangle + \overline{\beta}\langle x, z\rangle$$

for all $x, y, z \in V$ and $\alpha, \beta \in \mathbb{K}$.

Example 1.40 Let $\mathbb{K}^n$ be the vector space of all ordered n-tuples $(x_1, x_2, \ldots, x_n)$, where $x_i \in \mathbb{K}$ for $i \in \{1, 2, \ldots, n\}$. We define a mapping $\langle\cdot,\cdot\rangle : \mathbb{K}^n \times \mathbb{K}^n \to \mathbb{K}$ by

$$\langle x, y\rangle = \sum_{i=1}^{n} x_i \overline{y}_i$$

for all ordered n-tuples $x = (x_1, x_2, \ldots, x_n)$ and $y = (y_1, y_2, \ldots, y_n)$. Then $\langle\cdot,\cdot\rangle$ is an inner product for the vector space $\mathbb{K}^n$ and $(\mathbb{K}^n, \langle\cdot,\cdot\rangle)$ is an inner product space.

Every inner product $\langle \cdot, \cdot \rangle$ for a vector space V naturally induces an associated norm through $\|x\| = \sqrt{\langle x, x \rangle}$ for all $x \in V$. Therefore, each inner product space is a normed space. In general, the norm on an inner product space means the mapping defined by $\|x\| = \sqrt{\langle x, x \rangle}$.

Theorem 1.41 (Cauchy–Schwarz Inequality) *Let* $(V, \langle \cdot, \cdot \rangle)$ *be an inner product space over* $\mathbb{K}$. *Then*

$$|\langle x, y \rangle| \leq \|x\| \|y\|$$

for all $x, y \in V$.

Proof For any $x, y \in V$, we set $A = \|x\|^2 = \langle x, x \rangle$, $B = |\langle x, y \rangle|$, and $C = \|y\|^2 = \langle y, y \rangle$. We note that B is a nonnegative real number. Then we can choose a scalar $\alpha \in \mathbb{K}$ such that $|\alpha| = 1$ and $B = \alpha \langle y, x \rangle$. For all $r \in \mathbb{R}$, it follows from Definition 1.39 (iii) that

$$\begin{aligned} 0 &\leq \langle x - r\alpha y, x - r\alpha y \rangle \\ &= \langle x, x \rangle - r\alpha \langle y, x \rangle - r\overline{\alpha} \langle x, y \rangle + r^2 \langle y, y \rangle \\ &= \langle x, x \rangle - r\alpha \langle y, x \rangle - r\overline{\alpha \langle y, x \rangle} + r^2 \langle y, y \rangle \\ &= A - Br - \overline{B}r + Cr^2 \\ &= A - 2Br + Cr^2. \end{aligned}$$

If $C = 0$, it has to be $B = 0$. (Otherwise, the aforementioned inequality does not hold for large $r > 0$.) If $C > 0$, we take $r = \frac{B}{C}$ in the above inequality and we obtain $B^2 \leq AC$, which completes the proof. □

We may now ask whether every normed space is an inner product space. However, contrary to our expectation, the answer is negative. In the following theorem, we propose a necessary and sufficient condition for a normed space to be an inner product space.

We recall that the norm on an inner product space is defined by $\|x\| = \sqrt{\langle x, x \rangle}$.

Theorem 1.42 (Parallelogram Law) *Let* $(V, \langle \cdot, \cdot \rangle)$ *be an inner product space. Then*

$$\|x + y\|^2 + \|x - y\|^2 = 2\|x\|^2 + 2\|y\|^2 \tag{1.1}$$

for all $x, y \in V$.

Proof In view of Definition 1.39 (iii), we have

$$\begin{aligned}\|x+y\|^2 &= \langle x+y, x+y\rangle \\ &= \langle x,x\rangle + \langle x,y\rangle + \langle y,x\rangle + \langle y,y\rangle \\ &= \|x\|^2 + \langle x,y\rangle + \langle y,x\rangle + \|y\|^2\end{aligned}$$

for all $x, y \in V$. If we replace y by $-y$ in the above equation, then we obtain

$$\|x-y\|^2 = \|x\|^2 - \langle x,y\rangle - \langle y,x\rangle + \|y\|^2$$

for all $x, y \in V$. Finally, we obtain the parallelogram law by adding the last two equations.

□

Assume that $(V, \|\cdot\|)$ is a normed space over $\mathbb{K}$. For any norm that satisfies the parallelogram law (1.1), the inner product that generates the norm is unique as a consequence of *polarization identity*. In the case of $\mathbb{K} = \mathbb{R}$, the polarization identity is given by

$$\langle x, y\rangle = \frac{1}{4}\left(\|x+y\|^2 - \|x-y\|^2\right) \tag{1.2}$$

for all $x, y \in V$. For the case of $\mathbb{K} = \mathbb{C}$, the polarization identity is given by

$$\langle x, y\rangle = \frac{1}{4}\left(\|x+y\|^2 - \|x-y\|^2\right) + \frac{i}{4}\left(\|ix-y\|^2 - \|ix+y\|^2\right) \tag{1.3}$$

for all $x, y \in V$.

Hence, if the parallelogram law is satisfied in a normed space $(V, \|\cdot\|)$, then the normed space is an inner product space that is correspondingly equipped with the inner product (1.2) or (1.3).

Using the Cauchy–Schwarz inequality and the inverse triangle inequality, we will prove that the inner product and the norm are continuous mappings.

Theorem 1.43 *Let $(V, \langle\cdot,\cdot\rangle)$ be an inner product space over $\mathbb{K}$. For any fixed $x, y \in V$, the mappings $\langle x, \cdot\rangle : V \to \mathbb{K}$, $\langle \cdot, y\rangle : V \to \mathbb{K}$, and $\|\cdot\| : V \to \mathbb{R}$ are (uniformly) continuous on V.*

Proof By the Cauchy–Schwarz inequality, we have

$$|\langle x, y_1\rangle - \langle x, y_2\rangle| = |\langle x, y_1 - y_2\rangle| \le \|x\|\|y_1 - y_2\|$$

for all $y_1, y_2 \in V$ and for a fixed $x \in V$. That is, $\langle x, \cdot\rangle$ is a uniformly continuous mapping. Similarly, we can prove that if y is a fixed element of V, then $\langle \cdot, y\rangle$ is a uniformly continuous mapping.

Furthermore, using the inverse triangle inequality, we obtain

$$\big|\|x_1\| - \|x_2\|\big| \le \|x_1 - x_2\|$$

for all $x_1, x_2 \in V$. Therefore, $\|\cdot\|$ is also a uniformly continuous mapping. □

We remember that any subset W of a vector space V is called a subspace of V if and only if W is itself a vector space with the same vector addition and scalar multiplication, which are defined on V. In sentences referring to vector spaces, the term "subspace" always refers to the subspace aforementioned.

Theorem 1.44 *If W is a subspace of an inner product space V, so is $\overline{W}$.*

Proof Let x and y be arbitrary elements of $\overline{W}$ and let α be an arbitrary scalar. Then there exist sequences $\{x_i\}$ and $\{y_i\}$ in W which converge to x and y, respectively. Since W is a vector space, $\{x_i + y_i\}$ and $\{\alpha x_i\}$ are sequences in W, which converge to $x + y$ and αx, respectively. Therefore, since $\overline{W}$ is closed, $x + y \in \overline{W}$ and $\alpha x \in \overline{W}$, which implies that $\overline{W}$ is a subspace of V by Definition 1.26. □

One of the most important uses of the inner product is to define the orthogonality of the vectors. This distinguishes the Hilbert space theory from the general theory of Banach spaces.

Definition 1.45 Let V be an inner product space. Two vectors $x, y \in V$ are called *orthogonal* if and only if $\langle x, y\rangle = 0$. In the case, we write $x \perp y$.

We notice that if $x \perp y$, then $\langle x, y\rangle = 0$, and hence, $\langle y, x\rangle = \overline{\langle x, y\rangle} = \overline{0} = 0$, i.e., $y \perp x$.

Another example of the geometric property of the norm defined by an inner product is the Pythagorean theorem. The Pythagorean theorem describes the basic relationship between the sides of a right triangle in terms of Euclidean geometry.

The following theorem shows that the Pythagorean theorem holds in each inner product space.

Theorem 1.46 (Pythagorean Theorem) *Let $(V, \langle\cdot,\cdot\rangle)$ be an inner product space. If the norm $\|\cdot\|$ on V is induced by the inner product $\langle\cdot,\cdot\rangle$, then the Pythagorean formula holds, i.e.,*

$$\|x + y\|^2 = \|x\|^2 + \|y\|^2$$

for any pair of orthogonal vectors in V.

Proof If x and y are orthogonal vectors, then $\langle x, y \rangle = \langle y, x \rangle = 0$. Hence, we have

$$\|x + y\|^2 = \|x\|^2 + \langle x, y \rangle + \langle y, x \rangle + \|y\|^2 = \|x\|^2 + \|y\|^2$$

for all $x, y \in V$. □

The Pythagorean theorem can be generalized into the following theorem.

Theorem 1.47 (Generalized Pythagorean Theorem) *Let V be an inner product space. If the norm of V is given as in Theorem 1.46, then the Pythagorean formula holds, i.e.,*

$$\left\| \sum_{i=1}^{n} x_i \right\|^2 = \sum_{i=1}^{n} \|x_i\|^2$$

for all orthogonal vectors $x_1, x_2, \ldots, x_n \in V$.

Proof We note that $\|x_1+x_2\|^2 = \|x_1\|^2+\|x_2\|^2$ holds for all orthogonal vectors $x_1, x_2 \in V$ according to Theorem 1.46. Hence, this theorem is true for $n = 2$. Assume now that this theorem holds for some integer $n > 1$. For any orthogonal vectors $x_1, x_2, \ldots, x_n, x_{n+1} \in V$, we set $x = \sum_{i=1}^{n} x_i$ and $y = x_{n+1}$. Since x and y are orthogonal, we have

$$\left\| \sum_{i=1}^{n+1} x_i \right\|^2 = \|x + y\|^2 = \|x\|^2 + \|y\|^2 = \sum_{i=1}^{n} \|x_i\|^2 + \|x_{n+1}\|^2 = \sum_{i=1}^{n+1} \|x_i\|^2,$$

which is our equality for $n + 1$. We can complete the proof with the inductive conclusion. □

We say that a set of vectors forms an *orthonormal set* if and only if all vectors in the set are mutually orthogonal and all have unit length. The Pythagorean formula will help us prove Bessel's inequality, which is an important property of orthonormal sets.

Theorem 1.48 (Bessel's Inequality) *Let $\{x_1, x_2, \ldots, x_n\}$ be an orthonormal set of vectors in an inner product space V. Then*

$$\sum_{i=1}^{n} |\langle x, x_i \rangle|^2 \leq \|x\|^2 \tag{1.4}$$

for all $x \in V$.

Proof It follows from Theorem 1.47 that

$$\left\| \sum_{i=1}^{n} c_i x_i \right\|^2 = \sum_{i=1}^{n} \|c_i x_i\|^2 = \sum_{i=1}^{n} |c_i|^2$$

for any scalars $c_1, c_2, \ldots, c_n$. Thus, we have

$$\begin{aligned}
\left\| x - \sum_{i=1}^{n} c_i x_i \right\|^2 &= \left\langle x - \sum_{i=1}^{n} c_i x_i,\ x - \sum_{j=1}^{n} c_j x_j \right\rangle \\
&= \|x\|^2 - \left\langle x,\ \sum_{j=1}^{n} c_j x_j \right\rangle - \left\langle \sum_{i=1}^{n} c_i x_i,\ x \right\rangle + \sum_{i=1}^{n} |c_i|^2 \|x_i\|^2 \\
&= \|x\|^2 - \sum_{j=1}^{n} \overline{c_j} \langle x, x_j \rangle - \sum_{i=1}^{n} c_i \overline{\langle x, x_i \rangle} + \sum_{i=1}^{n} c_i \overline{c_i} \\
&= \|x\|^2 - \sum_{i=1}^{n} |\langle x, x_i \rangle|^2 + \sum_{i=1}^{n} |\langle x, x_i \rangle - c_i|^2.
\end{aligned}$$

Since $c_1, c_2, \ldots, c_n$ are arbitrary scalars, if we set $c_i = \langle x, x_i \rangle$ for $i \in \{1, 2, \ldots, n\}$, then the previous equality yields

$$\left\| x - \sum_{i=1}^{n} \langle x, x_i \rangle x_i \right\|^2 = \|x\|^2 - \sum_{i=1}^{n} |\langle x, x_i \rangle|^2 \tag{1.5}$$

for all $x \in V$, which gives the Bessel's inequality (1.4). □

1.5 Hilbert Spaces

Since Hilbert spaces allow the methods of linear algebra and calculus to be generalized from (finite-dimensional) Euclidean spaces to possibly infinite-dimensional spaces, the Hilbert spaces are widely used in mathematics and physics.

Definition 1.49 A *Hilbert space* is a complete inner product space.

In the aforementioned definition, the completeness of an inner product space $(V, \langle \cdot, \cdot \rangle)$ means the completeness of the normed space $(V, \| \cdot \|)$, where the norm is defined by $\|x\| = \sqrt{\langle x, x \rangle}$ for all $x \in V$.

There are many examples of Hilbert spaces. $\mathbb{R}^n$ and $\mathbb{C}^n$ are Hilbert spaces if they are equipped with the inner products $\langle x, y \rangle = \sum_{i=1}^{n} x_i y_i$ and $\langle x, y \rangle = \sum_{i=1}^{n} x_i \overline{y}_i$, respectively. Another example of Hilbert spaces is ℓ^2, where ℓ^2 is the space of all sequences $\{x_i\}$ of real numbers such that $\sum_{i=1}^{\infty} x_i^2 < \infty$ with the inner product defined by

$$\langle x, y\rangle = \sum_{i=1}^{\infty} x_i y_i.$$

Two vectors in an inner product space are said to be *orthonormal* if and only if they are orthogonal unit vectors. We say that a set of vectors forms an *orthonormal set* if and only if all vectors in the set are mutually orthogonal and all have unit length.

Definition 1.50 Let $(V, \langle\cdot,\cdot\rangle)$ be an inner product space.

(i) A collection S of nonzero vectors in V is said to be an *orthogonal system* if and only if any two different vectors in S are orthogonal to each other.
(ii) An orthogonal system S is said to be an *orthonormal system* if and only if every vector in S is a unit vector, i.e., $\|x\| = 1$ for every $x \in S$.

In Definition 1.50 (ii), we note that the norm is induced by the inner product, i.e., it is defined by $\|x\| = \sqrt{\langle x, x\rangle}$.

Every orthogonal system of nonzero vectors can be normalized. If S is an orthogonal system, then the collection

$$S' = \left\{ \frac{1}{\|x\|} x : x \in S \right\}$$

is an orthonormal system.

Theorem 1.51 *Every orthogonal system in an inner product space is linearly independent.*

Proof Assume that S is an orthogonal system in an inner product space $(V, \langle\cdot,\cdot\rangle)$ over $\mathbb{K}$ and that $\sum_{i=1}^{n} \alpha_i x_i = 0$ for some $x_1, x_2, \ldots, x_n \in S$ and $\alpha_1, \alpha_2, \ldots, \alpha_n \in \mathbb{K}$. Then, we have

$$\begin{aligned} 0 &= \sum_{i=1}^{n} \langle 0, \alpha_i x_i\rangle = \sum_{i=1}^{n} \left\langle \sum_{j=1}^{n} \alpha_j x_j, \alpha_i x_i \right\rangle = \sum_{i=1}^{n} \langle \alpha_i x_i, \alpha_i x_i\rangle \\ &= \sum_{i=1}^{n} |\alpha_i|^2 \|x_i\|^2, \end{aligned}$$

which implies that $\alpha_i = 0$ for all $i \in \{1, 2, \ldots, n\}$. Therefore, $x_1, x_2, \ldots, x_n$ are linearly independent. □

Definition 1.52 Let V be an inner product space.

(i) A sequence of vectors of V that constitutes an orthonormal system is called an *orthonormal sequence*.

(ii) An orthonormal sequence $\{x_i\}$ in V is called *complete* if and only if

$$x = \sum_{i=1}^{\infty} \langle x, x_i \rangle x_i$$

for all $x \in V$.

We note that the equality of Definition 1.52 (ii) means

$$\lim_{n \to \infty} \left\| x - \sum_{i=1}^{n} \langle x, x_i \rangle x_i \right\| = 0,$$

where $\| \cdot \|$ is the norm on V defined by $\|v\| = \sqrt{\langle v, v \rangle}$ for any $v \in V$.

Lemma 1.53 *Assume that V is a Hilbert space over $\mathbb{K}$, $\{x_i\}$ is an orthonormal sequence in V, and that $\{c_i\}$ is a sequence of scalars from $\mathbb{K}$. Then the infinite series $\sum_{i=1}^{\infty} c_i x_i$ converges if and only if $\sum_{i=1}^{\infty} |c_i|^2 < \infty$. In that case,*

$$\left\| \sum_{i=1}^{\infty} c_i x_i \right\|^2 = \sum_{i=1}^{\infty} |c_i|^2.$$

Proof For any integers $n > m > 0$, it follows from Theorem 1.47 that

$$\left\| \sum_{i=m}^{n} c_i x_i \right\|^2 = \sum_{i=m}^{n} |c_i|^2. \tag{1.6}$$

If $\sum_{i=1}^{\infty} |c_i|^2 < \infty$, it follows from (1.6) that the sequence $\{s_n\}$, $s_n = \sum_{i=1}^{n} c_i x_i$, is a Cauchy sequence in V. This fact, together with the completeness of V, implies the convergence of the infinite series $\sum_{i=1}^{\infty} c_i x_i$.

Conversely, we assume that the infinite series $\sum_{i=1}^{\infty} c_i x_i$ converges. Since the sequence $\{\sigma_n\}$ of scalars from $\mathbb{K}$, where $\sigma_n = \sum_{i=1}^{n} |c_i|^2$, is a Cauchy sequence in $\mathbb{R}$, the convergence of $\sum_{i=1}^{\infty} |c_i|^2$ follows from (1.6).

Finally, it is enough to take $m = 1$ and put $n \to \infty$ in (1.6) to obtain our equality. □

Definition 1.54 An orthonormal system S in an inner product space V over $\mathbb{K}$ is called an *orthonormal basis* if and only if every element $x \in V$ has a unique representation

$$x = \sum_{i=1}^{\infty} c_i x_i,$$

where $c_i \in \mathbb{K}$ and x_i's are distinct elements of S.

We note that a complete orthonormal sequence is an orthonormal basis. In view of Definition 1.52 (ii), it is sufficient to show the uniqueness of the representation. We leave this work to the reader's practice.

Remark 1.55 Let V be an inner product space over $\mathbb{K}$. If $\{x_i\}$ is a complete orthonormal sequence in V, then the span

$$\operatorname{span}\{x_1, x_2, \ldots\} = \left\{ \sum_{i=1}^{n} c_i x_i : c_1, c_2, \ldots, c_n \in \mathbb{K};\ n \in \mathbb{N} \right\}$$

is dense in V.

Theorem 1.56 *Let V be a Hilbert space. An orthonormal sequence $\{x_i\}$ in V is complete if and only if $\langle x, x_i \rangle = 0$ for all $i \in \mathbb{N}$ implies $x = 0$.*

Proof Assume that $\{x_i\}$ is a complete orthonormal sequence in V. Then we have

$$x = \sum_{i=1}^{\infty} \langle x, x_i \rangle x_i$$

for all $x \in V$. Hence, if $\langle x, x_i \rangle = 0$ for all $i \in \mathbb{N}$, then $x = 0$.

Conversely, assume that $\langle x, x_i \rangle = 0$ for all $i \in \mathbb{N}$ implies $x = 0$. For any $x \in V$, we define

$$y = \sum_{i=1}^{\infty} \langle x, x_i \rangle x_i.$$

In view of Bessel's inequality, we have

$$\sum_{i=1}^{n} |\langle x, x_i \rangle|^2 \leq \|x\|^2$$

for any $x \in V$ and $n \in \mathbb{N}$. By letting $n \to \infty$ in Bessel's inequality, we get

$$\sum_{i=1}^{\infty} |\langle x, x_i \rangle|^2 \le \|x\|^2.$$

By Lemma 1.53 and the previous inequality, we conclude that the sum y exists.

We note that for any $i \in \mathbb{N}$,

$$\begin{aligned}
\langle x - y, x_i \rangle &= \langle x, x_i \rangle - \left\langle \sum_{k=1}^{\infty} \langle x, x_k \rangle x_k, x_i \right\rangle \\
&= \langle x, x_i \rangle - \sum_{k=1}^{\infty} \langle x, x_k \rangle \langle x_k, x_i \rangle \\
&= \langle x, x_i \rangle - \langle x, x_i \rangle \\
&= 0,
\end{aligned}$$

which implies that $x = y = \sum_{i=1}^{\infty} \langle x, x_i \rangle x_i$. According to Definition 1.52 (ii), the orthonormal sequence $\{x_i\}$ is complete. □

The *Parseval's formula* presented in the following theorem can be interpreted as an extension of the Pythagorean formula to infinite sums.

Theorem 1.57 (Parseval's Formula) *An orthonormal sequence $\{x_i\}$ in a Hilbert space V is complete if and only if*

$$\|x\|^2 = \sum_{i=1}^{\infty} |\langle x, x_i \rangle|^2 \tag{1.7}$$

for all $x \in V$.

Proof Assume that x is any element of V. If $\{x_i\}$ is a complete orthonormal sequence, then the expression on the left hand in (1.5) converges to 0 as $n \to \infty$. Thus, we have

$$\lim_{n \to \infty} \left(\|x\|^2 - \sum_{i=1}^{n} |\langle x, x_i \rangle|^2 \right) = 0,$$

which implies the validity of (1.7).

Conversely, if the Parseval's formula (1.7) holds, then the expression on the right hand in (1.5) converges to 0 as $n \to \infty$. Hence, we obtain

$$\lim_{n\to\infty}\left\|x-\sum_{i=1}^{n}\langle x, x_i\rangle x_i\right\|^2=0,$$

which implies that the orthonormal sequence $\{x_i\}$ is complete. □

Definition 1.58 A Hilbert space V_1 is said to be *isomorphic* to a Hilbert space V_2 if and only if there exists a bijective linear mapping $f : V_1 \to V_2$ such that

$$\langle f(x), f(y)\rangle = \langle x, y\rangle$$

for all $x, y \in V_1$. In this case, the mapping f is called a *Hilbert space isomorphism* from V_1 onto V_2.

We note that a Hilbert space is said to be *separable* if and only if it contains a complete orthonormal sequence (see Definition 1.9 or [16]). It should be noted that all finite-dimensional Hilbert spaces are separable.

Theorem 1.59 *If V is a real Hilbert space with* $\dim V = n$*, then it is isomorphic to* $\mathbb{R}^n$.

Proof Considering the Gram–Schmidt orthonormalization, we can assume that $\{x_1, x_2, \ldots, x_n\}$ is an orthonormal basis of V and x is an arbitrary element of V. We define a mapping $f : V \to \mathbb{R}^n$ by $f(x) = (c_1, c_2, \ldots, c_n)$, where $c_i = \langle x, x_i\rangle$ for $i \in \{1, 2, \ldots, n\}$. Then it is easy to check that f is a one-to-one linear mapping from V onto $\mathbb{R}^n$.

In addition, for arbitrary $x, y \in V$, we set $c_i = \langle x, x_i\rangle$ and $d_i = \langle y, x_i\rangle$, $i \in \{1, 2, \ldots, n\}$. Then we have

$$\begin{aligned}
\langle f(x), f(y)\rangle &= \big\langle (c_1, c_2, \ldots, c_n), (d_1, d_2, \ldots, d_n)\big\rangle \\
&= \sum_{i=1}^{n} c_i d_i = \sum_{i=1}^{n}\langle x, x_i\rangle\langle y, x_i\rangle \\
&= \sum_{i=1}^{n}\big\langle x, \langle y, x_i\rangle x_i\big\rangle \\
&= \left\langle x, \sum_{i=1}^{n}\langle y, x_i\rangle x_i\right\rangle \\
&= \langle x, y\rangle.
\end{aligned}$$

Therefore, according to Definition 1.58, f is an isomorphism from V onto $\mathbb{R}^n$. □

We note that any real infinite-dimensional separable Hilbert space is isomorphic to the real space ℓ^2, which is presented in the next section. In this sense, there is only one real infinite-dimensional separable Hilbert space. The same applies to complex Hilbert spaces: any two complex infinite-dimensional separable Hilbert spaces are isomorphic.

1.6 Hilbert Space ℓ^2

Let $\mathbb{R}^\infty$ be the set of all real-valued sequences. We define

$$\{x_i\} + \{y_i\} = \{x_i + y_i\} \quad \text{and} \quad \alpha\{x_i\} = \{\alpha x_i\}$$

for all $\{x_i\}, \{y_i\} \in \mathbb{R}^\infty$ and $\alpha \in \mathbb{R}$. Moreover, we define

$$-\{x_i\} = \{-x_i\}$$

for any $\{x_i\} \in \mathbb{R}^\infty$. Using these definitions as the vector addition and the scalar multiplication, we can easily check that $\mathbb{R}^\infty$ is a real vector space.

We now define

$$\ell^2 = \left\{ \{x_i\} \in \mathbb{R}^\infty : \sum_{i=1}^{\infty} x_i^2 < \infty \right\}$$

as a subset of $\mathbb{R}^\infty$, consisting of all real-valued sequences $\{x_i\}$ with the property $\sum_{i=1}^{\infty} x_i^2 < \infty$.

Theorem 1.60 *Assume that* $\{x_i\}, \{y_i\} \in \ell^2$. *Then* $\sum_{i=1}^{\infty} x_i y_i$ *converges and*

$$\left| \sum_{i=1}^{\infty} x_i y_i \right| \le \left(\sum_{i=1}^{\infty} x_i^2 \right)^{1/2} \left(\sum_{i=1}^{\infty} y_i^2 \right)^{1/2}.$$

Proof For each $n \in \mathbb{N}$, the ordered n-tuple $(x_1, x_2, \ldots, x_n)$ is an element of $\mathbb{R}^n$, where $\mathbb{R}^n$ is the n-dimensional Euclidean space. Thus, by using the Cauchy-Schwarz inequality for $\mathbb{R}^n$, we obtain

$$\sum_{i=1}^{n} |x_i||y_i| = |\langle x, y \rangle_n| \le \|x\|_n \|y\|_n = \left(\sum_{i=1}^{n} x_i^2 \right)^{1/2} \left(\sum_{i=1}^{n} y_i^2 \right)^{1/2},$$

where we define

$$\langle x, y\rangle_n = \sum_{i=1}^{n} |x_i||y_i| \quad \text{and} \quad \|x\|_n = \left(\sum_{i=1}^{n} x_i^2\right)^{1/2}$$

for all elements $x = (x_1, x_2, \ldots, x_n)$ and $y = (y_1, y_2, \ldots, y_n)$ of $\mathbb{R}^n$. Since $\{x_i\}, \{y_i\} \in \ell^2$, we have

$$\sum_{i=1}^{n} |x_i||y_i| \le \left(\sum_{i=1}^{\infty} x_i^2\right)^{1/2} \left(\sum_{i=1}^{\infty} y_i^2\right)^{1/2}.$$

Hence, the finite sum of the left-hand side is bounded above, and it increases with n. Therefore, we have

$$\sum_{i=1}^{\infty} |x_i||y_i| \le \left(\sum_{i=1}^{\infty} x_i^2\right)^{1/2} \left(\sum_{i=1}^{\infty} y_i^2\right)^{1/2}.$$

That is, $\sum_{i=1}^{\infty} x_i y_i$ converges absolutely and thus, it converges.

Moreover, we obtain

$$\left|\sum_{i=1}^{\infty} x_i y_i\right| \le \sum_{i=1}^{\infty} |x_i||y_i| \le \left(\sum_{i=1}^{\infty} x_i^2\right)^{1/2} \left(\sum_{i=1}^{\infty} y_i^2\right)^{1/2}$$

for all $\{x_i\}, \{y_i\} \in \ell^2$. □

Using the previous theorem, we can now show that ℓ^2 is a subspace of the real vector space $\mathbb{R}^\infty$.

Theorem 1.61 *ℓ^2 is a subspace of the real vector space $\mathbb{R}^\infty$.*

Proof We just need to show that ℓ^2 is closed under the action of vector addition and scalar multiplication. It is obvious that if $x \in \ell^2$, then $\alpha x \in \ell^2$ for any $\alpha \in \mathbb{R}$.

Assume that $x = \{x_i\} \in \ell^2$ and $y = \{y_i\} \in \ell^2$. It then follows from Theorem 1.60 that $\sum_{i=1}^{\infty} x_i y_i$ converges, since both $\sum_{i=1}^{\infty} x_i^2$ and $\sum_{i=1}^{\infty} x_i^2$ converge. We note that

$$\sum_{i=1}^{n} (x_i + y_i)^2 = \sum_{i=1}^{n} x_i^2 + 2\sum_{i=1}^{n} x_i y_i + \sum_{i=1}^{n} y_i^2$$

for every $n \in \mathbb{N}$. Since every sequence of partial sums on the right-hand side converges, the sequence on the left-hand side also converges. That is, $\sum_{i=1}^{\infty}(x_i + y_i)^2$ converges, which implies that $x + y \in \ell^2$. □

Now we define a real-valued mapping $\langle \cdot, \cdot \rangle : \ell^2 \times \ell^2 \to \mathbb{R}$ by

$$\langle x, y \rangle = \sum_{i=1}^{\infty} x_i y_i \tag{1.8}$$

for all $x = \{x_i\} \in \ell^2$ and $y = \{y_i\} \in \ell^2$.

Theorem 1.62 $\langle \cdot, \cdot \rangle$ *is an inner product for* ℓ^2.

Proof It is not difficult to check that the given mapping $\langle \cdot, \cdot \rangle$ satisfies the four conditions given in Definition 1.39. We therefore leave the proof of this theorem to the reader. □

Theorem 1.60 represents the ℓ^2-version of the Cauchy–Schwarz inequality. The inner product defined by (1.8) induces the metric d on ℓ^2:

$$d(x, y) = \sqrt{\langle x - y, x - y \rangle} = \left(\sum_{i=1}^{\infty} (x_i - y_i)^2 \right)^{1/2}. \tag{1.9}$$

We also note that ℓ^2 contains the $\tilde{\mathbb{R}}^n$ for each $n \in \mathbb{N}$, where $\tilde{\mathbb{R}}^n$ is an isomorphic copy of $\mathbb{R}^n$. More precisely, we define

$$\tilde{\mathbb{R}}^n = \big\{ \{x_i\} \in \mathbb{R}^\infty : x_i = 0 \text{ for all } i > n \big\}.$$

Thus, the metric on ℓ^2 given by (1.9) can be regarded as a generalization of the Euclidean metric.

Theorem 1.63 ℓ^2 *is complete.*

Proof Assume that $\{q_m\}$ is an arbitrary Cauchy sequence in ℓ^2, where we set

$$q_m = \{q_{m,1}, q_{m,2}, \ldots, q_{m,n}, \ldots\}$$

for every $m \in \mathbb{N}$.

(a) We assert that for each $j \in \mathbb{N}$, $\{q_{m,j}\}_m$ is a Cauchy sequence in $\mathbb{R}$. Let j be fixed and let $\varepsilon > 0$ be arbitrarily given. Then there exists an integer $N > 0$ such that

$d(q_m, q_n) < \varepsilon$ for all $m, n > N$, where the metric d on ℓ^2 is defined by (1.9). Moreover, we have

$$|q_{m,j} - q_{n,j}| \leq \left(\sum_{i=1}^{\infty} (q_{m,i} - q_{n,i})^2 \right)^{1/2} = d(q_m, q_n) < \varepsilon$$

for all $m, n > N$. Therefore, for every $j \in \mathbb{N}$, $\{q_{m,j}\}_m$ is a Cauchy sequence in $\mathbb{R}$ and there exists a real number $z_j = \lim_{m \to \infty} q_{m,j}$. We now let $z = \{z_j\}$.

(*b*) We check that $z \in \ell^2$. Let $h > 0$ be an arbitrary real number. Since $\{q_m\}$ is a Cauchy sequence in ℓ^2, there is an integer $N > 0$ such that $d(q_N, q_{N+p}) < h$ for all integers $p > 0$. Then we have

$$\sum_{i=1}^{\infty} \left(q_{N,i} - q_{N+p,i}\right)^2 = d\left(q_N, q_{N+p}\right)^2 < h^2$$

and hence, we obtain

$$\sum_{i=1}^{n} \left(q_{N,i} - q_{N+p,i}\right)^2 < h^2$$

for every $n \in \mathbb{N}$. Since $\lim_{p \to \infty} q_{N+p,i} = z_i$ for any $i \in \mathbb{N}$, it follows from the previous inequality that

$$\lim_{p \to \infty} \sum_{i=1}^{n} \left(q_{N,i} - q_{N+p,i}\right)^2 = \sum_{i=1}^{n} \left(q_{N,i} - z_i\right)^2 \leq h^2.$$

Since the sequence of real numbers

$$\left\{ \sum_{i=1}^{n} \left(q_{N,i} - z_i\right)^2 \right\}_n$$

is monotone increasing and bounded above, it converges. More precisely,

$$\sum_{i=1}^{\infty} \left(q_{N,i} - z_i\right)^2 < \infty.$$

Therefore, we conclude that $\{q_{N,i} - z_i\}_i \in \ell^2$. We note that

$$\{z_i\} = \{q_{N,i}\}_i - \{q_{N,i} - z_i\}_i \in \ell^2,$$

since $\{q_{N,i}\}_i = q_N \in \ell^2$ and $\{q_{N,i} - z_i\}_i \in \ell^2$. That is, $z = \{z_i\} \in \ell^2$.

(c) We will prove that $\lim_{n\to\infty} q_n = z$. Given $\varepsilon > 0$, there exists an integer $N > 0$ such that $d(q_n, q_{n+p}) < \varepsilon$ for all integers $n > N$ and $p > 0$. Then we have

$$\sum_{i=1}^{\infty}(q_{n,i} - q_{n+p,i})^2 < \varepsilon^2$$

for all integers $n > N$ and $p > 0$. Hence, we further obtain

$$\sum_{i=1}^{k}(q_{n,i} - q_{n+p,i})^2 < \varepsilon^2$$

for any integer $k > 0$. Thus, since $\lim_{p\to\infty} q_{n+p,i} = z_i$ by (a), we have

$$\sum_{i=1}^{k}(q_{n,i} - z_i)^2 \le \varepsilon^2$$

for each integer $k > 0$. Hence,

$$\sum_{i=1}^{\infty}(q_{n,i} - z_i)^2 \le \varepsilon^2$$

for all integers $n > N$, which implies that $d(q_n, z) \le \varepsilon$ for each integer $n > N$. That is, the Cauchy sequence $\{q_n\}$ converges to $z \in \ell^2$ by (b). Therefore, every Cauchy sequence in ℓ^2 converges (in ℓ^2), i.e., ℓ^2 is complete. □

We have proved in the preceding theorems that $(\ell^2, \langle\cdot,\cdot\rangle)$ is a complete real inner product space, i.e., it is a real Hilbert space.

Theorem 1.64 *$(\ell^2, \langle\cdot,\cdot\rangle)$ is a real Hilbert space.*

An important idea for the proof of the following theorem is that for each $\varepsilon > 0$, every $x_0 \in \ell^2$ is within ε distance from the subset

$$D_n = \left\{\{x_i\} \in \ell^2 : x_i \in \mathbb{Q} \text{ for } 0 < i \le n \text{ and } x_i = 0 \text{ for } i > n\right\} \tag{1.10}$$

for some integer $n > 0$.

Theorem 1.65 *ℓ^2 is separable.*

Proof First, we easily check that D_n defined by (1.10) is a countable subset of ℓ^2. We now define

$$D = \bigcup_{n=1}^{\infty} D_n$$

and note that D is also a countable subset of ℓ^2.

We assert that D is dense in ℓ^2. Assume that $p = \{p_i\}$ is arbitrary element of ℓ^2 and $\varepsilon > 0$ is given arbitrarily. We complete the proof by showing that there is an element $z \in D$ that satisfies $d(z, p) < \varepsilon$. Since $p \in \ell^2$, we have $\sum_{i=1}^{\infty} p_i^2 < \infty$. Thus, we can choose an integer $N > 0$ such that

$$\sum_{i=N+1}^{\infty} p_i^2 < \frac{1}{2}\varepsilon^2.$$

Furthermore, we may choose rational numbers $r_1, r_2, \ldots, r_N$ such that

$$|r_i - p_i| < \frac{1}{\sqrt{2N}}\varepsilon$$

for $i \in \{1, 2, \ldots, N\}$. If we set $z = \{r_1, r_2, \ldots, r_N, 0, 0, \ldots\}$, then $z \in D$ and we obtain

$$\begin{aligned} d(z, p) &= \left(\sum_{i=1}^{N} (r_i - p_i)^2 + \sum_{i=N+1}^{\infty} p_i^2 \right)^{1/2} \\ &< \left(\frac{N}{2N}\varepsilon^2 + \frac{1}{2}\varepsilon^2 \right)^{1/2} \\ &= \varepsilon, \end{aligned}$$

which implies that the countable subset D (of ℓ^2) is dense in ℓ^2. Therefore, ℓ^2 is separable.

□

2 Aleksandrov Problem

Abstract

Isometry is not only widely used to describe the motion of rigid bodies that is a mixture of rotational and translational motions of solids but is also an important tool in theoretical physics and geometry. The properties of isometry have been actively researched for a long time, and this theory is widely used in fields such as natural science, computer engineering, radiology, imaging science, and design. In this chapter, we will examine in detail the Aleksandrov problem, one of the ways to find necessary and sufficient conditions for characterizing the isometries defined between Euclidean spaces.

2.1 Theorem of Mazur and Ulam

Let $(X, \|\cdot\|)$ and $(Y, \|\cdot\|)$ be some normed spaces. A mapping $f : X \to Y$ is called an *isometry* if and only if f satisfies the equality $\|f(x) - f(y)\| = \|x - y\|$ for all $x, y \in X$. In other words, a mapping is an isometry if and only if it preserves all distances.

Definition 2.1 Let $f : X \to Y$ be a mapping between the real normed spaces X and Y.

(i) f is called *linear* if and only if it satisfies $f(sx+ty) = sf(x)+tf(y)$ for all $x, y \in X$ and $s, t \in \mathbb{R}$.

(ii) f is *affine* if and only if it satisfies $f((1-t)x + ty) = (1-t)f(x) + tf(y)$ for all $x, y \in X$ and $t \in \mathbb{R}$.

The original version of the chapter has been revised. A correction to this chapter can be found at https://doi.org/10.1007/978-3-031-77613-7_9

S.-M. Jung, *Aleksandrov-Rassias Problems on Distance Preserving Mappings*, Frontiers in Mathematics, https://doi.org/10.1007/978-3-031-77613-7_2

Remark 2.2 Combining (i) and (ii) of Definition 2.1, the mapping f is affine if and only if it is linear up to a translation, and this happens if and only if $f - f(0)$ is linear.

Proof Assume that the mapping $f : X \to Y$ is affine. Let us define the mapping $g : X \to Y$ by $g(x) = f(x) - f(0)$ for all $x \in X$. It then follows from Definition 2.1 (ii) that

$$g((1-t)x + ty) = (1-t)g(x) + tg(y) \tag{2.1}$$

for all $x, y \in X$ and $t \in \mathbb{R}$. If we set $x = 0$ in (2.1), then we obtain

$$g(ty) = tg(y) \tag{2.2}$$

for any $y \in X$ and $t \in \mathbb{R}$. Moreover, we set $x = \frac{1}{2}$ in (2.1) and we use (2.2) to have

$$g(x + y) = g(x) + g(y) \tag{2.3}$$

for all $x, y \in X$.

We note that

$$sx + ty = \frac{s}{t}(tx + ty) + \left(1 - \frac{s}{t}\right)(ty)$$

for all $x, y \in X$ and $s, t \in \mathbb{R}$. Thus, by (2.1), (2.2), and (2.3), we get

$$\begin{aligned} g(sx + ty) &= g\left(\frac{s}{t}(tx + ty) + \left(1 - \frac{s}{t}\right)(ty)\right) \\ &= \frac{s}{t}g(tx + ty) + \left(1 - \frac{s}{t}\right)g(ty) \\ &= sg(x + y) + (t - s)g(y) \\ &= sg(x) + tg(y) \end{aligned}$$

for all $x, y \in X$ and $s, t \in \mathbb{R}$. That is, g is linear.

The reverse implication is obviously true. □

The theory of isometries in Banach spaces originates from the paper [50] by S. Mazur and S. M. Ulam. They proved the theorem named after them, which states that every surjective isometry between real normed spaces is affine.

Theorem 2.3 (Mazur and Ulam) *Let X and Y be real normed spaces. Every surjective isometry $f : X \to Y$ is affine.*

Proof We assume that $f : X \to Y$ is an arbitrary surjective isometry between real normed spaces X and Y. We will show that f preserves midpoints, i.e.,

$$f\left(\frac{1}{2}(x+y)\right) = \frac{1}{2}\big(f(x) + f(y)\big) \tag{2.4}$$

for all $x, y \in X$.

(a) First, we assume that g is an arbitrary surjective isometry defined on X. Tentatively, let x and y be two fixed vectors of X. We define

$$\operatorname{def}(g) = \left\| g\left(\frac{1}{2}(x+y)\right) - \frac{1}{2}\big(g(x) + g(y)\big) \right\|,$$

which denotes the possible "affine defect." Using the triangle inequality and our assumption that g is an isometry, we have

$$\begin{aligned}
\operatorname{def}(g) &\le \frac{1}{2}\left\| g\left(\frac{1}{2}(x+y)\right) - g(x) \right\| + \frac{1}{2}\left\| g\left(\frac{1}{2}(x+y)\right) - g(y) \right\| \\
&= \frac{1}{2}\left\| \frac{1}{2}(x+y) - x \right\| + \frac{1}{2}\left\| \frac{1}{2}(x+y) - y \right\| \\
&= \frac{1}{2}\|x - y\|
\end{aligned}$$

for any surjective isometry g defined on X. Since x, y are tentatively fixed vectors, the affine defect $\operatorname{def}(\cdot)$ has a uniform bound.

(b) We claim that for every surjective isometry g defined on X, there is another surjective isometry h (defined on X) whose affine defect is twice as large as the affine defect of g. In fact, the one-to-one correspondence of g allows us to define $h = g^{-1} \circ r \circ g$, where r is the reflection in $\frac{1}{2}(g(x) + g(y))$ in the range space of g. Hence, the reflection r is given by $r(z) = g(x) + g(y) - z$, and $h(x) = y$, $h(y) = x$. Since g^{-1} is also an isometry, we obtain

$$\begin{aligned}
\operatorname{def}(h) &= \left\| h\left(\frac{1}{2}(x+y)\right) - \frac{1}{2}\big(h(x) + h(y)\big) \right\| \\
&= \left\| g^{-1}\left(g(x) + g(y) - g\left(\frac{1}{2}(x+y)\right)\right) - \frac{1}{2}(x+y) \right\| \\
&= \left\| g^{-1}\left(g(x) + g(y) - g\left(\frac{1}{2}(x+y)\right)\right) - g^{-1}\left(g\left(\frac{1}{2}(x+y)\right)\right) \right\| \\
&= \left\| g(x) + g(y) - g\left(\frac{1}{2}(x+y)\right) - g\left(\frac{1}{2}(x+y)\right) \right\| \\
&= 2\operatorname{def}(g).
\end{aligned}$$

(c) If we had a surjective isometry defined on X with positive affine defect, then the repeated use of part (b) leads to the conclusion that we would obtain surjective isometries (defined on X) with arbitrarily large affine defect, which would contradict the uniform boundedness of the affine defect that we established in (a). Thus, we conclude that $\mathrm{def}(g) = 0$ for every surjective isometry g defined on X.

(d) Therefore, the surjective isometry $f : X \to Y$ satisfies $\mathrm{def}(f) = 0$, i.e., f satisfies the Eq. (2.4) for all $x, y \in X$. (For convenience, x and y are assumed to be tentatively fixed, but in practice, they can be chosen arbitrarily.)

(e) Since every isometry is obviously a continuous mapping, we now only have to prove that all continuous solutions of Eq. (2.4) are affine mappings. Equation (2.4) is called the "Jensen's functional equation" and every solution to this equation is called a "Jensen function." It is well known that each mapping f between real vector spaces with $f(0) = 0$ is a Jensen function if and only if it is an additive mapping. Moreover, any continuous additive mapping is linear. Therefore, according to [31, §7.1] or [41, Result 1.54], we can conclude that $f - f(0)$ is linear, i.e., f is affine. □

The aforementioned proof of the Mazur–Ulam theorem is an improved version of a new proof presented by B. Nica [52]. It should be noted that, prior to B. Nica, J. Väisälä presented another new proof of the Mazur–Ulam theorem (see [70]).

2.2 Theorem of Baker

As we see in Sect. 2.1, Mazur and Ulam proved that every isometry from one real normed space "onto" another is necessarily affine. It is natural to ask whether the Mazur–Ulam theorem is also true without the "onto" assumption.

In connection with this question, we introduce an important property of special normed spaces: A normed space is *strictly convex* if and only if $\|x + y\| = \|x\| + \|y\|$ implies that x and y are linearly dependent. It is easy to see that a normed space is strictly convex if and only if $x \neq 0$, $y \neq 0$, and $\|x + y\| = \|x\| + \|y\|$ imply $x = cy$ for some $c > 0$. A detailed definition of strict convexity is introduced as follows.

Definition 2.4 A normed space $(X, \|\cdot\|)$ is *strictly convex* if and only if for any nonzero vectors $x, y \in X$ that satisfy $\|x+y\| = \|x\|+\|y\|$, there exists a constant $c > 0$, depending on x and y, such that $x = cy$.

For every strictly convex normed space X, we can check that if x and y are distinct unit vectors of X, then $\|\alpha x + (1 - \alpha)y\| < 1$ for all scalars α satisfying $0 < \alpha < 1$: Using the triangle inequality, we have

$$\|\alpha x + (1 - \alpha)y\| \leq \|\alpha x\| + \|(1 - \alpha)y\| = \alpha\|x\| + (1 - \alpha)\|y\| = 1.$$

On the contrary, assume that $\|\alpha x + (1-\alpha)y\| = 1$ for some $0 < \alpha < 1$. Since

$$\begin{aligned}\|\alpha x + (1-\alpha)y\| = 1 \\ &= \alpha + (1-\alpha) \\ &= \alpha\|x\| + (1-\alpha)\|y\| \\ &= \|\alpha x\| + \|(1-\alpha)y\|.\end{aligned}$$

On account of the strict convexity of X, there exists a constant $c > 0$ such that $\alpha x = c(1-\alpha)y$ or $x = c\frac{1-\alpha}{\alpha}y$. Since x and y are unit vectors, it follows that $x = y$, which contradicts our assumption that $x \neq y$. Therefore, we can conclude that $\|\alpha x + (1-\alpha)y\| < 1$ for all real numbers α with $0 < \alpha < 1$.

As a typical example of the strictly convex space, we can take the real Hilbert space we are familiar with.

Theorem 2.5 *Every real Hilbert space is strictly convex.*

Proof Let x and y be arbitrary nonzero elements of the real Hilbert space X with $\|x + y\| = \|x\| + \|y\|$. Since $\|x\| = \sqrt{\langle x, x\rangle}$ for all $x \in X$, it follows from Definition 1.39 (iii) that

$$\|x+y\|^2 = \langle x+y, x+y\rangle = \|x\|^2 + 2\langle x, y\rangle + \|y\|^2$$

and

$$\big(\|x\| + \|y\|\big)^2 = \|x\|^2 + 2\|x\|\|y\| + \|y\|^2.$$

Equating the last two equations, we get

$$\langle x, y\rangle = \|x\|\|y\|.$$

Furthermore, using the Cauchy–Schwarz inequality, we have

$$\|x\|\|y\| = \langle x, y\rangle \leq |\langle x, y\rangle| \leq \|x\|\|y\|,$$

i.e., $|\langle x, y\rangle| = \|x\|\|y\|$, which implies that x and y are linearly dependent. Hence, there exists a real constant $c \neq 0$ such that $x = cy$. It follows from $\|x+y\| = \|x\| + \|y\|$ that $|c+1| = |c| + 1$, which yields that $c > 0$. Due to Definition 2.4, we can conclude that the real Hilbert space X is strictly convex. In general, every real Hilbert space is strictly convex. □

Lemma 2.6 *Let Y be a real normed space that is strictly convex. If $x, y \in Y$, then $\frac{1}{2}(x+y)$ is the unique element of Y separated from both x and y by the distance $\frac{1}{2}\|x-y\|$.*

Proof The claim is obviously true if $x = y$. It is also easy to see that the element $\frac{1}{2}(x+y)$ is separated from both x and y by a distance $\frac{1}{2}\|x-y\|$. Therefore, all that remains is to prove the uniqueness.

Assume that $x \neq y$ and $u, v \in Y$ with

$$\|x-u\| = \|y-u\| = \|x-v\| = \|y-v\| = \frac{1}{2}\|x-y\|.$$

Then we have

$$\begin{aligned} \left\|x - \frac{1}{2}(u+v)\right\| &= \left\|\frac{1}{2}(x-u) + \frac{1}{2}(x-v)\right\| \\ &\leq \frac{1}{2}\|x-u\| + \frac{1}{2}\|x-v\| \\ &= \frac{1}{2}\|x-y\|. \end{aligned} \tag{2.5}$$

Similarly, we obtain

$$\left\|y - \frac{1}{2}(u+v)\right\| \leq \frac{1}{2}\|x-y\|. \tag{2.6}$$

If either of these inequalities were strict, then we would have

$$\|x-y\| \leq \left\|x - \frac{1}{2}(u+v)\right\| + \left\|y - \frac{1}{2}(u+v)\right\| < \|x-y\|,$$

a contradiction. Therefore, equality in (2.5) and (2.6) must hold, so that

$$\left\|\frac{1}{2}(x-u) + \frac{1}{2}(x-v)\right\| = \frac{1}{2}\|x-y\| = \left\|\frac{1}{2}(x-u)\right\| + \left\|\frac{1}{2}(x-v)\right\|.$$

Since Y is strictly convex, $x \neq u$ and $x \neq v$, it follows that $x-u = c(x-v)$ for some real number $c > 0$. However, since $\|x-u\| = \|x-v\|$, we have $c = 1$ and thus $u = v$. □

J. A. Baker [2] significantly generalized the Mazur–Ulam theorem by proving the following theorem without the "onto" condition:

Theorem 2.7 (Baker) *If X is a real normed space and Y is a strictly convex real normed space, then every isometry $f : X \to Y$ is affine.*

Proof If $f(0) \neq 0$, then we define a mapping $g : X \to Y$ by $g(x) = f(x) - f(0)$. Then, g is also an isometry and $g(0) = 0$. Therefore, without loss of generality, we can assume that $f(0) = 0$.

Since f is an isometry, it follows that $\|f(x)\| = \|x\|$ for all $x \in X$. Hence, $\|-f(-x)\| = \|f(-x)\| = \|-x\| = \|x\|$ for each $x \in X$. Moreover, we have

$$\big\|f(x) + (-f(-x))\big\| = 2\|x\| = \|f(x)\| + \|-f(-x)\|.$$

Since Y is strictly convex, there is a $c > 0$ such that $f(x) = -cf(-x)$. But it is true that $\|f(x)\| = \|x\| = \|f(-x)\|$ and it follows that $c = 1$. Thus, it is true that $f(-x) = -f(x)$ for all $x \in X$.

Since f is an isometry and $f(0) = 0$, we obtain

$$\begin{aligned}\|f(x+y)\| &= \|x+y\| \\ &= \|x - (-y)\| \\ &= \|f(x) - f(-y)\| \\ &= \|f(x) + f(y)\|\end{aligned}$$

for all $x, y \in X$. Furthermore, we have

$$\left\|f\left(\frac{1}{2}(x+y)\right) - f(x)\right\| = \left\|\frac{1}{2}(y-x)\right\| = \frac{1}{2}\|x-y\| = \frac{1}{2}\|f(x) - f(y)\|$$

and similarly,

$$\left\|f\left(\frac{1}{2}(x+y)\right) - f(y)\right\| = \frac{1}{2}\|f(x) - f(y)\|$$

for all $x, y \in X$.

It follows from Lemma 2.6 that

$$f\left(\frac{1}{2}(x+y)\right) = \frac{1}{2}\big(f(x) + f(y)\big)$$

for any $x, y \in X$. Every solution to this equation is called a Jensen function. Any mapping f between real vector spaces with $f(0) = 0$ is a Jensen function if and only if it is additive. Moreover, any continuous additive mapping is linear. Since f is continuous as an isometry, it is linear and the proof is complete. □

The following simple but interesting lemma was presented in the paper [20] by P. Fischer and Gy. Muszély:

Lemma 2.8 *If Y is a normed space, $x, y \in Y$, and $\|x+y\| = \|x\|+\|y\|$, then $\|sx+ty\| = s\|x\| + t\|y\|$ for all $s, t \geq 0$.*

Proof If we assume that $0 \leq s \leq t$, then we have $\|sx + ty\| \leq s\|x\| + t\|y\|$. On the other hand, using the inverse triangle inequality, we obtain

$$\begin{aligned}\|sx + ty\| &= \|t(x + y) - (t - s)x\| \\ &\geq \big|t\|x + y\| - (t - s)\|x\|\big| \\ &= s\|x\| + t\|y\|,\end{aligned}$$

which completes our proof. □

Let Y be a real normed space that is not strictly convex. In this case, we can see from the following example that we have an isometry that is not affine.

Example 2.9 Assume that Y is a real normed space that is not strictly convex. Due to Definition 2.4 and Lemma 2.8, we can choose $a, b \in Y$ such that a and b are linearly independent, $\|a\| = \|b\| = 1$, and $\|a + b\| = \|a\| + \|b\|$.

Now we define a mapping $f : \mathbb{R} \to Y$ by

$$f(x) = \begin{cases} xa & (\text{for } x \leq 1), \\ a + (x - 1)b & (\text{for } x > 1). \end{cases}$$

We note that

$$f(y) - f(x) = \begin{cases} (y - x)a & (\text{for } x \leq y \leq 1), \\ (1 - x)a + (y - 1)b & (\text{for } x \leq 1 < y), \\ (y - x)b & (\text{for } 1 < x \leq y). \end{cases}$$

Using Lemma 2.8, we can easily show that f is an isometry. Furthermore, it holds that $f(2) = a + b \neq 2a = 2f(1)$. Therefore, f is neither linear nor affine.

As another example, we are going to construct a homogeneous isometry that is not linear.

Example 2.10 Let us define a mapping $g : \mathbb{R}^2 \to \mathbb{R}$ by

$$g(x, y) = \begin{cases} x \ (\text{for } xy \geq 0 \text{ and } |x| \leq |y|), \\ y \ (\text{for } xy \geq 0 \text{ and } |y| \leq |x|), \\ 0 \ (\text{for } xy < 0). \end{cases}$$

It is easy to check that

(i) g is homogeneous, i.e., $g(tx, ty) = tg(x, y)$ for all $t, x, y \in \mathbb{R}$;
(ii) $|g(x, y) - g(u, v)| \leq \sqrt{(x-u)^2 + (y-v)^2}$ for all $u, v, x, y \in \mathbb{R}$;
(iii) g is not linear.

For example, the inequality (ii) can be checked by considering the following nine cases. Due to space constraints, we will discuss the first three and last of the nine cases.

(1) If the points x, y, u, v in $\mathbb{R}$ satisfy the conditions that $xy \geq 0$, $|x| \leq |y|$, $uv \geq 0$, and $|u| \leq |v|$, then

$$|g(x, y) - g(u, v)| = |x - u| \leq \sqrt{(x-u)^2 + (y-v)^2}.$$

(2) If the points x, y, u, v in $\mathbb{R}$ satisfy the conditions that $xy \geq 0$, $|x| \leq |y|$, $uv \geq 0$, and $|v| \leq |u|$, then

$$\begin{aligned} |g(x, y) - g(u, v)| &= |x - v| \\ &\leq \max\left\{|x-u|, |y-v|\right\} \\ &\leq \sqrt{(x-u)^2 + (y-v)^2}. \end{aligned}$$

(3) If the points x, y, u, v in $\mathbb{R}$ satisfy the conditions that $xy \geq 0$, $|x| \leq |y|$, and $uv < 0$, then

$$\begin{aligned} |g(x, y) - g(u, v)| &= |x - 0| \\ &\leq \max\left\{|x-u|, |y-v|\right\} \\ &\leq \sqrt{(x-u)^2 + (y-v)^2}. \end{aligned}$$

In this way, we calculate up to the last case and check:

(9) If the points x, y, u, v in $\mathbb{R}$ satisfy the conditions that $xy < 0$ and $uv < 0$, then

$$|g(x, y) - g(u, v)| = |0 - 0| \leq \sqrt{(x-u)^2 + (y-v)^2}.$$

We set $X = \mathbb{R}^2$ with the usual normed space structure and $Y = \mathbb{R}^3$ with the usual vector space structure but with

$$\|(x, y, z)\| = \max\left\{\sqrt{x^2 + y^2}, |z|\right\}.$$

Then, we can check that $\|\cdot\| : Y \to [0, \infty)$ is a norm on Y.

If we define $f : X \to Y$ by $f(x, y) = (x, y, g(x, y))$, then we have

$$\begin{aligned}\|f(x, y) - f(u, v)\| &= \left\|\big(x - u, y - v, g(x, y) - g(u, v)\big)\right\| \\ &= \max\left\{\sqrt{(x - u)^2 + (y - v)^2}, |g(x, y) - g(u, v)|\right\} \\ &= \sqrt{(x - u)^2 + (y - v)^2} \\ &= \|(x, y) - (u, v)\|\end{aligned}$$

for all $(x, y), (u, v) \in \mathbb{R}^2$, which implies that f is an isometry. Finally, it follows from (i), (ii), and (iii) that f is a homogeneous isometry but not linear.

2.3 Aleksandrov Problem

Many mathematicians have long studied the methods of characterizing isometries, one of which is what is known as the *Aleksandrov problem*, and it is a central goal of this book to explain this problem in detail so that readers can grasp it.

Definition 2.11 Let $f : X \to Y$ be a mapping between normed spaces.

(i) A distance $\rho > 0$ is called *contractive* (or *non-expanding*) by f if and only if $\|x - y\| = \rho$ always implies $\|f(x) - f(y)\| \le \rho$.

(ii) A distance $\rho > 0$ is called *extensive* (or *non-shrinking*) by f if and only if the inequality $\|f(x) - f(y)\| \ge \rho$ holds for all $x, y \in X$ with $\|x - y\| = \rho$.

(iii) A distance $\rho > 0$ is called *preserved* (or *conservative*) by f if and only if ρ is both contractive and extensive by f.

Remark 2.12 Let $f : X \to Y$ be a mapping between normed spaces, and let n be any positive integer. If f preserves a distance ρ, then the distance $n\rho$ is contractive.

Proof Let x and y be arbitrary elements of X that satisfies $\|x - y\| = n\rho$ and let $f : X \to Y$ be a mapping that preserves the distance ρ. We define

$$x_i = x + \frac{i}{n}(y - x)$$

for each $i \in \{0, 1, \ldots, n\}$. Then we have $x_0 = x$ and $x_n = y$. Moreover, we obtain

$$\|x_i - x_{i+1}\| = \frac{1}{n}\|x - y\| = \rho$$

for any $i \in \{0, 1, \ldots, n-1\}$. Since the distance ρ is preserved by f, it follows that

$$\|f(x_i) - f(x_{i+1})\| = \rho$$

for any $i \in \{0, 1, \ldots, n-1\}$. Furthermore, using the triangle inequality, we have

$$\begin{aligned}\|f(x) - f(y)\| &= \|f(x_0) - f(x_n)\| \\ &\leq \sum_{i=0}^{n-1} \|f(x_i) - f(x_{i+1})\| \\ &= n\rho.\end{aligned}$$

That is, the distance $n\rho$ is contractive by f. □

If f is an isometry, then f preserves all distances $\rho > 0$, and *vice versa*. At this point, we may raise a question:

Is a mapping that preserves certain distances an isometry?

In relation to this question, A. D. Aleksandrov [1] posed the following problem.

Problem 2.13 (Aleksandrov Problem) Does the existence of a single conservative distance for a mapping imply that it is an isometry?

It is now known as the *Aleksandrov problem.*

The following examples show that the Aleksandrov problem is nontrivial by presenting that there are some mappings that may not be isometries even if they preserve a certain distance.

Example 2.14 We define the mapping $f : \mathbb{R} \to \mathbb{R}$ by

$$f(x) = \begin{cases} x + 1 & (\text{for } x \in \mathbb{Z}), \\ x & (\text{for } x \notin \mathbb{Z}). \end{cases}$$

If $x, y \in \mathbb{R}$ with $|x - y| = 1$, then either $|f(x) - f(y)| = |(x + 1) - (y + 1)| = 1$ for all integers x and y or $|f(x) - f(y)| = |x - y| = 1$ for all non-integers x and y. That is, f preserves the unit distance. However, if $x = \frac{3}{2}$ and $y = 0$, then $|x - y| = \frac{3}{2}$ but $|f(x) - f(y)| = |\frac{3}{2} - 1| = \frac{1}{2}$, which shows that $|f(x) - f(y)| \neq |x - y|$. That is, f is not an isometry.

Example 2.15 There is a mapping $f : \ell^2 \to \ell^2$ that is not an isometry, even if it preserves the unit distance: Since ℓ^2 is separable by Theorem 1.65, there exists a countable subset D of ℓ^2 that is dense in ℓ^2, where we set $D = \{d_1, d_2, \ldots\}$. We choose a mapping $g : \ell^2 \to D$ such that $\|g(x) - x\| < \frac{1}{2}$ for all $x \in \ell^2$, where $\|x\|^2 = \langle x, x\rangle = \sum_{i=1}^{\infty} x_i^2$. Now, we define a mapping $h : D \to \ell^2$ by

$$h(d_i) = \left\{ \frac{1}{\sqrt{2}}\delta_{i1}, \frac{1}{\sqrt{2}}\delta_{i2}, \ldots, \frac{1}{\sqrt{2}}\delta_{ii}, \ldots \right\}$$

for each $i \in \mathbb{N}$, where δ_{ij} is the Kronecker delta.

We assert that the mapping $f = h \circ g : \ell^2 \to \ell^2$ preserves the unit distance: Assume that x_1 and x_2 are arbitrary points of ℓ^2 with $\|x_1 - x_2\| = 1$. By the inverse triangle inequality, we have

$$\begin{aligned} \|g(x_1) - g(x_2)\| &= \big\| \big(g(x_1) - x_1\big) - \big(g(x_2) - x_2\big) - (x_2 - x_1) \big\| \\ &\geq \Big| \big\| \big(g(x_1) - x_1\big) - \big(g(x_2) - x_2\big) \big\| - \|x_2 - x_1\| \Big| \\ &= 1 - \big\| \big(g(x_1) - x_1\big) - \big(g(x_2) - x_2\big) \big\| \\ &> 0, \end{aligned}$$

since $\|(g(x_1) - x_1) - (g(x_2) - x_2)\| \leq \|g(x_1) - x_1\| + \|g(x_2) - x_2\| < 1$. Hence, $g(x_1) \neq g(x_2)$. Since $g(x_1), g(x_2) \in D$ and $g(x_1) \neq g(x_2)$, there are two distinct $m, n \in \mathbb{N}$ such that $d_m = g(x_1)$ and $d_n = g(x_2)$. Thus $f(x_1)$ is the real-valued sequence whose mth term is $\frac{1}{\sqrt{2}}$ and whose remaining terms are all 0s. In the same way, we see that $f(x_2)$ is the real-valued sequence whose nth term is $\frac{1}{\sqrt{2}}$ and whose remaining terms are all 0s. Hence, $f(x_1) \neq f(x_2)$. Therefore, we have

$$\|f(x_1) - f(x_2)\|^2 = \left(\frac{1}{\sqrt{2}}\right)^2 + \left(-\frac{1}{\sqrt{2}}\right)^2 = 1.$$

We assert that not all distances are preserved by f: By the argument of the preceding paragraph, we see that $\|f(x_1) - f(x_2)\| \in \{0, 1\}$ for all $x_1, x_2 \in \ell^2$. That is, f is not an isometry.

It is not yet known how the aforementioned example would change if the mappings involved were continuous. Another example of the mapping that preserves unit distance, but is not an isometry, was introduced by Th. M. Rassias [56]:

Example 2.16 We define a continuous mapping $f : \mathbb{R} \to \mathbb{R}$ by

$$f(x) = [x] + \{x\}^2,$$

where $[x]$ denotes the integer part of x and $\{x\} = x - [x]$. Since $[x + 1] = [x] + 1$ and $\{x + 1\} = x + 1 - [x + 1] = x - [x] = \{x\}$, we have

$$f(x + 1) = [x + 1] + \{x + 1\}^2 = [x] + 1 + \{x\}^2 = f(x) + 1$$

for any $x \in \mathbb{R}$. That is, f preserves the unit distance. But we see that $f(0) = 0$ and $f(\frac{1}{2}) = \frac{1}{4}$. Therefore, f is not an isometry.

Although a mapping $f : \mathbb{R} \to \mathbb{R}$ is continuous and preserves unit distance, Example 2.16 shows that f may not be an isometry. However, under the Lipschitz condition with a Lipschitz constant 1, which is a stronger concept than continuity, we get a different result. The following theorem, which was proved in the paper [60] by Rassias and Xiang, is a pedagogically interesting example.

Theorem 2.17 (Rassias and Xiang) *Suppose $f : \mathbb{R} \to \mathbb{R}$ is a Lipschitz mapping with a Lipschitz constant 1, i.e.,*

$$|f(x) - f(y)| \leq |x - y| \tag{2.7}$$

for all $x, y \in \mathbb{R}$. If f preserves the unit distance, then f is an affine isometry.

Proof Without loss of generality, assume that $f(0) = 0$. Otherwise, we substitute $f(x) - f(0)$ for $f(x)$ for all $x \in \mathbb{R}$. Since f preserves the unit distance, it follows that $|f(1) - f(0)| = 1$. Hence, we have either $f(1) = 1$ or $f(1) = -1$.

(*a*) Assume that $f(1) = 1$. We will prove by induction on n that $f(n) = n$ for all integers $n \geq 0$. Suppose $f(m) = m$ for any $m \in \{0, 1, \ldots, n - 1\}$, where $n > 1$ is an integer. Since $|f(n) - f(n-1)| = |f(n) - (n-1)| = 1$, we obtain $f(n) = n-2$ or $f(n) = n$. Assume that $f(n) = n - 2$. Let $r = \frac{1}{2}((n - 2) + (n - 1))$. It then follows from (2.7) and the inverse triangle inequality that

$$\begin{aligned}
\frac{1}{2} &\geq |f(r) - f(n-2)| \\
&\geq \big||f(r) - f(n-1)| - |f(n-2) - f(n-1)|\big| \\
&\geq 1 - |r - (n-1)| \\
&= \frac{1}{2},
\end{aligned}$$

which implies that $|f(r) - f(n - 2)| = \frac{1}{2}$. Applying the same argument, it follows that $|f(r) - f(n - 1)| = \frac{1}{2}$. Since $f(n - 2) = n - 2$ and $f(n - 1) = n - 1$, we have

$$f(r) = \frac{1}{2}\big((n-2) + (n-1)\big).$$

We set $s = \frac{1}{2}((n-1) + n)$. Similarly, we obtain

$$\begin{aligned}\frac{1}{2} &\geq |f(s) - f(n-1)| \\ &\geq \big||f(s) - f(n)| - |f(n-1) - f(n)|\big| \\ &\geq 1 - |s-n| \\ &= \frac{1}{2},\end{aligned}$$

which implies that $|f(s) - f(n-1)| = \frac{1}{2}$. Using the same argument, we get $|f(s) - f(n)| = \frac{1}{2}$. Since $f(n-1) = n-1$ and $f(n) = n-2$, we further have

$$f(s) = \frac{1}{2}\big((n-2) + (n-1)\big).$$

Therefore, $|f(r) - f(s)| = 0$ and $|r - s| = 1$, which contradicts the assumption that f preserves the unit distance. Hence, $f(n) = n$ for every integer $n \geq 0$.

(b) Since f preserves the unit distance and $f(0) = 0$, it follows that either $f(-1) = -1$ or $f(-1) = 1$. Assume that $f(-1) = 1$. It follows from (2.7) that

$$\left|1 - f\left(-\frac{1}{2}\right)\right| = \left|f(-1) - f\left(-\frac{1}{2}\right)\right| \leq \frac{1}{2}$$

and

$$\left|1 - f\left(\frac{1}{2}\right)\right| = \left|f(1) - f\left(\frac{1}{2}\right)\right| \leq \frac{1}{2}.$$

Moreover, by (2.7), we have

$$\left|f\left(-\frac{1}{2}\right)\right| \leq \frac{1}{2} \text{ and } \left|f\left(\frac{1}{2}\right)\right| \leq \frac{1}{2}.$$

From the last four inequalities, it follows that $f(-\frac{1}{2}) = f(\frac{1}{2}) = \frac{1}{2}$, which contradicts our assumption that f preserves the unit distance. Hence, we can conclude that $f(-1) = -1$.

(c) We now define the mapping $g : \mathbb{R} \to \mathbb{R}$ by $g(x) = -f(-x)$ for every $x \in \mathbb{R}$. Then, g also satisfies condition (2.7) and g preserves unit distance if and only if f does. Furthermore, $g(1) = 1$ by (b). Hence, it follows from (a) that $g(n) = n$ for all

$n \in \mathbb{N}_0$, which is equivalent to the conclusion that $f(n) = n$ for all integers $n \le 0$. Therefore, we can conclude by considering (a) that $f(n) = n$ for all $n \in \mathbb{Z}$.

(d) For every $x \in \mathbb{R}$ there is an integer n_0 such that $n_0 \le x \le n_0 + 1$. Since $f : \mathbb{R} \to \mathbb{R}$ satisfies inequality (2.7), we obtain

$$|f(x) - f(n_0)| \le |x - n_0| \quad \text{and} \quad |f(x) - f(n_0 + 1)| \le |x - (n_0 + 1)|.$$

Taking into account (c), i.e., the fact that $f(n) = n$ for all $n \in \mathbb{Z}$, we can transform the above inequalities into

$$2n_0 - x \le f(x) \le x \quad \text{and} \quad x \le f(x) \le 2(n_0 + 1) - x.$$

We can easily solve the last inequalities as follows: $f(x) = x$. That is, $f(x) = x$ for all $x \in \mathbb{R}$. Therefore, f is a linear isometry.

(e) Finally, we examine the case $f(1) = -1$. Define $h(x) = -f(x)$ for all $x \in \mathbb{R}$. Then, $h(1) = 1$ and h preserves unit distance and satisfies condition (2.7). According to (d), h must be a linear isometry and $h(x) = x$ for all $x \in \mathbb{R}$. Therefore, f is a linear isometry and $f(x) = -x$ for all $x \in \mathbb{R}$. □

2.4 Theorem of Beckman and Quarles

As we have seen in the preceding section, Aleksandrov asked in 1970 whether the existence of a single conservative distance for a mapping implies that it is an isometry. However, it is quite interesting that F. S. Beckman and D. A. Quarles [3] partially solved the Aleksandrov problem in 1953, when the Aleksandrov problem had not yet been raised. For this reason we might have used the term "Aleksandrov-Beckman-Quarles problem" instead of the term "Aleksandrov problem." However, in this book, we will continue to refer to this problem as the Aleksandrov problem, following the convention.

A *Euclidean space* is a finite-dimensional inner product space over $\mathbb{R}$. From now on, we use $\mathbb{E}^n$ and $|| \cdot ||$ to denote the n-dimensional Euclidean space and its corresponding *Euclidean norm*, respectively. The Euclidean inner product $\langle \cdot, \cdot \rangle$ on $\mathbb{E}^n$ is defined as

$$\langle x, y \rangle = \sum_{i=1}^{n} x_i y_i$$

for all points $x = (x_1, x_2, \ldots, x_n)$ and $y = (y_1, y_2, \ldots, y_n)$ of $\mathbb{E}^n$.

Although "affine mapping" and "linear mapping up to a translation" are synonymous, we will prefer to use "affine mapping" from now on.

Theorem 2.18 (Beckman and Quarles) *Let n be a fixed integer greater than 1. If a mapping $f : \mathbb{E}^n \to \mathbb{E}^n$ preserves a distance $\rho > 0$, then f is an affine isometry.*

Remark 2.19 If a mapping $f : X \to Y$ between normed spaces preserves a distance $\rho > 0$, we can then assume that $\rho = 1$ (see [58]): For example, we define another mapping $g : X \to Y$ by $g(x) = \frac{1}{\rho} f(\rho x)$ for all $x \in X$. If f preserves the distance ρ, then g preserves the distance 1. On the other hand, if g preserves the unit distance, then

$$\|f(x) - f(y)\| = \left\|\rho g\left(\frac{1}{\rho}x\right) - \rho g\left(\frac{1}{\rho}y\right)\right\| = \rho\left\|\frac{1}{\rho}x - \frac{1}{\rho}y\right\| = \|x - y\|$$

for all $x, y \in X$ with $\|x - y\| = \rho$. That is, f preserves the distance ρ. For this reason, we often assume that the mapping f preserves the distance 1, instead of assuming that the mapping f preserves the distance ρ.

Beckman and Quarles [3] actually proved their theorem with respect to multi-valued mappings, but we can easily replace their original theorem with Theorem 2.18. Since the proof presented by Beckman and Quarles contains parts that are difficult to understand, the relatively easy-to-understand proof by W. Benz is here presented instead of the original authors' proof (see [5]). Throughout this book, "distance 1" and "unit distance" are understood synonymously.

Lemma 2.20 *Let a, b, m be arbitrary points in $\mathbb{E}^n$, where n is an integer greater than 1. Then, $\|m - a\| = \|b - m\| = \frac{1}{2}\|b - a\|$ if and only if $m = \frac{1}{2}(a + b)$.*

Proof In view of Theorem 2.5, it only needs to replace Y, x, and y in Lemma 2.6 with $\mathbb{E}^n$, a, and b, respectively. □

Let n be a fixed integer greater than 1. Suppose we are given n points in $\mathbb{E}^n$. If the distances between any two points are all $\beta > 0$, then we call the set of these n points a *β-set*.

Lemma 2.21 *Assume that α and β are real positive numbers satisfying*

$$\gamma(\alpha, \beta) = 4\alpha^2 - 2\beta^2\left(1 - \frac{1}{n}\right) > 0.$$

If P is a β-set, then there are exactly two distinct points in $\mathbb{E}^n$ with distance α from each point in P. Those two points will be called the α-associated points of P. The distance between these α-associated points is $\sqrt{\gamma(\alpha, \beta)}$.

Proof

(a) Let $P = \{p_1, p_2, \ldots, p_n\}$ be a β-set. Then for $i, j \in \{1, 2, \ldots, n-1\}$ with $i \neq j$ we have

$$\langle p_i - p_n, p_j - p_n \rangle = \frac{1}{2}\beta^2, \tag{2.8}$$

since

$$\begin{aligned}\beta^2 &= \|p_i - p_j\|^2 \\ &= \|(p_i - p_n) - (p_j - p_n)\|^2 \\ &= \|p_i - p_n\|^2 - 2\langle p_i - p_n, p_j - p_n \rangle + \|p_j - p_n\|^2.\end{aligned}$$

We now define

$$\lambda_r = \frac{\beta}{\sqrt{2r(r+1)}}$$

for all $r \in \mathbb{N}$ and also define $e_1, e_2, \ldots, e_{n-1}$ inductively by using the formula

$$(1+s)\lambda_s e_s = (p_s - p_n) - \sum_{r=1}^{s-1} \lambda_r e_r \tag{2.9}$$

for $s \in \{1, 2, \ldots, n-1\}$. Obviously, $\langle e_1, e_1 \rangle = 1$. We prove that $\{e_1, e_2, \ldots, e_{n-1}\}$ is an orthonormal set in $\mathbb{E}^n$, i.e.,

$$\langle e_i, e_j \rangle = \begin{cases} 1 \text{ (for } i = j < n), \\ 0 \text{ (for } i < j < n) \end{cases} \tag{2.10}$$

by applying the following three steps. Indeed, the Eq. (2.10) will be proved in turn for (i, j) following the order of the sequence

$$(1, 1),\ (1, 2),\ (2, 2),\ (1, 3),\ (2, 3),\ (3, 3),\ (1, 4),\ \ldots,\ (n-1, n-1). \tag{2.11}$$

(i) Step $(i, i) \to (1, i+1)$: This step is applied to prove equation (2.10) when jumping from $(i, j) = (1, 1)$ to $(1, 2)$, or from $(2, 2)$ to $(1, 3)$, or from $(3, 3)$ to $(1, 4)$, ..., or from $(n-2, n-2)$ to $(1, n-1)$. It follows from (2.8) and (2.9) that

$$
\begin{aligned}
\frac{1}{2}\beta^2 &= \langle p_1 - p_n, p_{i+1} - p_n \rangle \\
&= \left\langle 2\lambda_1 e_1, (i+2)\lambda_{i+1} e_{i+1} + \sum_{r=1}^{i} \lambda_r e_r \right\rangle \\
&= 2(i+2)\lambda_1 \lambda_{i+1} \langle e_1, e_{i+1} \rangle + \sum_{r=1}^{i} 2\lambda_1 \lambda_r \langle e_1, e_r \rangle \\
&= 2(i+2)\lambda_1 \lambda_{i+1} \langle e_1, e_{i+1} \rangle + 2\lambda_1^2 \\
&= 2(i+2)\lambda_1 \lambda_{i+1} \langle e_1, e_{i+1} \rangle + \frac{1}{2}\beta^2,
\end{aligned}
$$

since $\langle e_1, e_1 \rangle = 1$ and $\langle e_1, e_2 \rangle = \langle e_1, e_3 \rangle = \cdots = \langle e_1, e_i \rangle = 0$ by inductive assumption. Hence, we obtain $\langle e_1, e_{i+1} \rangle = 0$.

(*ii*) Step $(i-1, i) \to (i, i)$: This step is applied to prove equation (2.10) when jumping from $(i, j) = (1, 2)$ to $(2, 2)$, or from $(2, 3)$ to $(3, 3)$, or from $(3, 4)$ to $(4, 4)$, …, or from $(n-2, n-1)$ to $(n-1, n-1)$. We obtain

$$
\begin{aligned}
\beta^2 &= \langle p_i - p_n, p_i - p_n \rangle \\
&= \left\langle \sum_{r=1}^{i-1} \lambda_r e_r + (1+i)\lambda_i e_i, \sum_{s=1}^{i-1} \lambda_s e_s + (1+i)\lambda_i e_i \right\rangle \\
&= \sum_{r=1}^{i-1} \sum_{s=1}^{i-1} \lambda_r \lambda_s \langle e_r, e_s \rangle + (1+i)\lambda_i \sum_{r=1}^{i-1} \lambda_r \langle e_r, e_i \rangle \\
&\quad + (1+i)\lambda_i \sum_{s=1}^{i-1} \lambda_s \langle e_i, e_s \rangle + (1+i)^2 \lambda_i^2 \langle e_i, e_i \rangle \\
&= \sum_{r=1}^{i-1} \lambda_r^2 + (1+i)^2 \lambda_i^2 \langle e_i, e_i \rangle,
\end{aligned}
$$

since $\langle e_r, e_s \rangle = 0$ for $r, s \in \{1, 2, \ldots, i-1\}$ with $r \neq s$, $\langle e_r, e_i \rangle = 0$ for $r \in \{1, 2, \ldots, i-1\}$, and $\langle e_i, e_s \rangle = 0$ for $s \in \{1, 2, \ldots, i-1\}$ by inductive assumption. Thus, we obtain $\langle e_i, e_i \rangle = 1$.

(*iii*) Step $(i-1, j) \to (i, j)$ for $i < j$: This step is applied to prove equation (2.10) when jumping from $(i, j) = (1, 3)$ to $(2, 3)$, or from $(1, 4)$ to $(2, 4)$, or from $(2, 4)$ to $(3, 4)$, …, or from $(n-3, n-1)$ to $(n-2, n-1)$. First, we remark that

$$\sum_{r=1}^{i-1} \lambda_r^2 + (1+i)\lambda_i^2 = \sum_{r=1}^{i-1} \frac{1}{2}\beta^2\left(\frac{1}{r} - \frac{1}{r+1}\right) + (1+i)\frac{1}{2}\beta^2 \frac{1}{i(i+1)} = \frac{1}{2}\beta^2.$$

It follows from (2.8) and (2.9) that

$$\begin{aligned}
\frac{1}{2}\beta^2 &= \langle p_i - p_n, p_j - p_n \rangle \\
&= \left\langle \sum_{r=1}^{i-1} \lambda_r e_r + (1+i)\lambda_i e_i, \sum_{s=1}^{j-1} \lambda_s e_s + (1+j)\lambda_j e_j \right\rangle \\
&= \sum_{r=1}^{i-1} \sum_{s=1}^{j-1} \lambda_r \lambda_s \langle e_r, e_s \rangle + (1+j)\lambda_j \sum_{r=1}^{i-1} \lambda_r \langle e_r, e_j \rangle \\
&\quad + (1+i)\lambda_i \sum_{s=1}^{j-1} \lambda_s \langle e_i, e_s \rangle + (1+i)\lambda_i (1+j)\lambda_j \langle e_i, e_j \rangle \\
&= \sum_{r=1}^{i-1} \lambda_r^2 + (1+i)\lambda_i^2 + (1+i)\lambda_i (1+j)\lambda_j \langle e_i, e_j \rangle,
\end{aligned}$$

and hence, we obtain $\langle e_i, e_j \rangle = 0$.

(*b*) We now assume that q is a point of $\mathbb{E}^n$ and the distance from each $p_s \in P$ is α, which implies that

$$\langle q - p_n, p_s - p_n \rangle = \frac{1}{2}\beta^2 \tag{2.12}$$

for all $s \in \{1, 2, \ldots, n-1\}$, since

$$\begin{aligned}
\alpha^2 &= \|q - p_s\|^2 \\
&= \|(q - p_n) - (p_s - p_n)\|^2 \\
&= \|q - p_n\|^2 - 2\langle q - p_n, p_s - p_n \rangle + \|p_s - p_n\|^2 \\
&= \alpha^2 - 2\langle q - p_n, p_s - p_n \rangle + \beta^2.
\end{aligned}$$

We set $q - p_n = \sum_{r=1}^{n} \mu_r e_r$, where $\mu_r \in \mathbb{R}$, by extending $\{e_1, e_2, \ldots, e_{n-1}\}$ of part (*a*) to an orthonormal basis $\{e_1, e_2, \ldots, e_{n-1}, e_n\}$ of $\mathbb{E}^n$. Then it follows from (2.9), (2.10), and (2.12) that

$$\frac{1}{2}\beta^2 = \langle q - p_n, p_s - p_n \rangle = \sum_{r=1}^{s-1} \mu_r \lambda_r + (1+s)\mu_s \lambda_s \tag{2.13}$$

for all $s \in \{1, 2, \dots, n-1\}$.

If we set $s = 1$ in (2.13), then we obtain $\mu_1 = \frac{1}{2}\beta = \lambda_1$. Assuming $\mu_i = \lambda_i$ for all $i \in \{1, 2, \dots, s-1\}$, where $s < n$, we also obtain $\mu_s = \lambda_s$ by comparing equation (2.13) with

$$\frac{1}{2}\beta^2 = \sum_{r=1}^{s-1} \lambda_r^2 + (1+s)\lambda_s^2.$$

Thus, $q - p_n = \sum_{r=1}^{n-1} \lambda_r e_r + \mu_n e_n$. Now, since $\alpha^2 = \|q - p_n\|^2 = \sum_{r=1}^{n-1} \lambda_r^2 + \mu_n^2$, we have

$$\mu_n^2 = \alpha^2 - \sum_{r=1}^{n-1} \lambda_r^2 = \alpha^2 - \frac{1}{2}\beta^2\left(1 - \frac{1}{n}\right) = \frac{1}{4}\gamma(\alpha, \beta).$$

There are exactly two solutions q, namely the points

$$q_i = p_n + \sum_{r=1}^{n-1} \lambda_r e_r \pm \frac{1}{2}\sqrt{\gamma(\alpha, \beta)}\, e_n, \tag{2.14}$$

where $i \in \{1, 2\}$, which are indeed of distance α from each point of the β-set P, i.e., they are the α-associated points of the β-set P. Obviously, we obtain $\|q_1 - q_2\| = \sqrt{\gamma(\alpha, \beta)}$. □

Lemma 2.22 *Assume that $\alpha, \beta > 0$ are real numbers satisfying $\gamma(\alpha, \beta) > 0$. Let x and y be arbitrary points in $\mathbb{E}^n$ separated by a distance $\|x - y\| = \sqrt{\gamma(\alpha, \beta)}$. Then there exists a β-set whose α-associated points are x and y.*

Proof Let us define

$$e_n = \frac{1}{\|y - x\|}(y - x) = \frac{1}{\sqrt{\gamma(\alpha, \beta)}}(y - x)$$

and extend $\{e_n\}$ to an orthonormal basis $\{e_1, e_2, \dots, e_n\}$ of $\mathbb{E}^n$. If p_n is a point of $\mathbb{E}^n$, then we inductively define $P = \{p_1, p_2, \dots, p_n\}$ by

$$p_s - p_n = \sum_{r=1}^{s-1} \lambda_r e_r + (1+s)\lambda_s e_s$$

for every $s \in \{1, 2, \ldots, n-1\}$, where $\lambda_r = \frac{\beta}{\sqrt{2r(r+1)}}$ for all $r \in \mathbb{N}$.

It is easy to check that P is a β-set. In precisely, we obtain

$$\|p_s - p_n\|^2 = \sum_{r=1}^{s-1} \lambda_r^2 + (1+s)^2\lambda_s^2 = \beta^2,$$

for all $s \in \{1, 2, \ldots, n-1\}$, and

$$\begin{aligned}
\|p_j - p_i\|^2 &= \|(p_j - p_n) - (p_i - p_n)\|^2 \\
&= \left\| \sum_{r=i+1}^{j-1} \lambda_r e_r + (1+j)\lambda_j e_j - i\lambda_i e_i \right\|^2 \\
&= \sum_{r=i+1}^{j-1} \lambda_r^2 + (1+j)^2\lambda_j^2 + (-i\lambda_i)^2 \\
&= \beta^2
\end{aligned}$$

for all integers i and j with $0 < i < j < n$.

If the point p_n is specially selected as

$$p_n = \frac{1}{2}(x+y) - \sum_{r=1}^{n-1} \lambda_r e_r,$$

then it follows from (2.14) that the α-associated points q_i of P are given by

$$q_i = p_n + \sum_{r=1}^{n-1} \lambda_r e_r \pm \frac{1}{2}\sqrt{\gamma(\alpha, \beta)}\, e_n = \frac{1}{2}(x+y) \pm \frac{1}{2}(y-x),$$

which implies that $\{q_1, q_2\} = \{x, y\}$. That is, x and y are the α-associated points of P. □

Proposition 2.23 *Let $\rho > 0$ be a fixed real number and let $n > 1$ and $N > 2$ be fixed integers. If a mapping $f : \mathbb{E}^n \to \mathbb{E}^n$ satisfies the following conditions*

(i) $\|x - y\| = \rho$ *implies* $\|f(x) - f(y)\| \le \rho$;
(ii) $\|x - y\| = N\rho$ *implies* $\|f(x) - f(y)\| = N\rho$

for all $x, y \in \mathbb{E}^n$, then f is an affine isometry.

Proof

(*a*) We will prove that f preserves the distances ρ and 2ρ. Let x and z be arbitrary points separated by a distance of 2ρ from each other. Then we can choose the midpoint $y = \frac{1}{2}(x + z)$ of x and z such that $\|x - y\| = \rho = \|y - z\|$. We put $p_\ell = x + \frac{\ell}{2}(z - x)$ for all $\ell \in \{0, 1, \ldots, N\}$. Then, it follows from the conditions (i) and (ii) that $\|f(p_0) - f(p_N)\| = N\rho$ and $\|f(p_\ell) - f(p_{\ell+1})\| \leq \rho$ for $\ell \in \{0, 1, \ldots, N - 1\}$ because of $\|p_0 - p_N\| = N\rho$ and $\|p_\ell - p_{\ell+1}\| = \rho$.

Moreover, using the triangle inequality, we obtain

$$\begin{aligned} N\rho &= \|f(p_0) - f(p_N)\| \\ &\leq \|f(p_0) - f(p_2)\| + \sum_{\ell=2}^{N-1} \|f(p_\ell) - f(p_{\ell+1})\| \\ &\leq \sum_{\ell=0}^{N-1} \|f(p_\ell) - f(p_{\ell+1})\| \\ &\leq N\rho, \end{aligned}$$

and hence, $\|f(p_\ell) - f(p_{\ell+1})\| = \rho$ for $\ell \in \{0, 1, \ldots, N - 1\}$ and

$$\|f(p_0) - f(p_2)\| = \|f(p_0) - f(p_1)\| + \|f(p_1) - f(p_2)\|.$$

We note that $p_0 = x$, $p_1 = y$, and $p_2 = z$. Considering the last equality, we can conclude that

$$\|f(x) - f(z)\| = 2\rho \quad \text{and} \quad \|f(x) - f(y)\| = \rho$$

for all $x, y, z \in \mathbb{E}^n$ with $\|x - z\| = 2\rho$ and $\|x - y\| = \rho$.

(*b*) We note that for all points x and y satisfying $\|x - y\| = \rho$, we can select a point $z = 2y - x$ so that $\|x - z\| = 2\rho$. Assume that x and y are arbitrary points of $\mathbb{E}^n$ and that the distance between these two points is ρ. We now assert that

$$f\big(x + \ell(y - x)\big) = f(x) + \ell\big(f(y) - f(x)\big) \tag{2.15}$$

for all $\ell \in \mathbb{N}_0$. If we put $p_\ell = x + \ell(y - x)$ for each $\ell \in \mathbb{N}_0$, then we obtain

$$\|p_\ell - p_{\ell-1}\| = \rho = \|p_{\ell+1} - p_\ell\| \quad \text{and} \quad \|p_{\ell+1} - p_{\ell-1}\| = 2\rho.$$

According to (a), f preserves the distances ρ and 2ρ. Thus, we have

$$\rho = \|f(p_\ell) - f(p_{\ell-1})\| = \|f(p_{\ell+1}) - f(p_\ell)\| = \frac{1}{2}\|f(p_{\ell+1}) - f(p_{\ell-1})\|,$$

and hence, by Lemma 2.20, we have

$$f(p_\ell) = \frac{1}{2}\big(f(p_{\ell-1}) + f(p_{\ell+1})\big) \quad \text{or} \quad f(p_{\ell+1}) = 2f(p_\ell) - f(p_{\ell-1}) \tag{2.16}$$

for each $\ell \in \mathbb{N}$.

Obviously, (2.15) is true for $\ell = 0$ or 1. Now we assume that (2.15) is true for all $\ell \in \{0, 1, \ldots, m\}$, where m is some positive integer. Then, by (2.15) and (2.16), we obtain

$$\begin{aligned} & f\big(x + (m+1)(y - x)\big) \\ & \quad = f(p_{m+1}) \\ & \quad = 2f(p_m) - f(p_{m-1}) \\ & \quad = 2\big(f(x) + m(f(y) - f(x))\big) - \big(f(x) + (m-1)(f(y) - f(x))\big) \\ & \quad = f(x) + (m+1)\big(f(y) - f(x)\big). \end{aligned}$$

By the mathematical induction, we conclude that (2.15) holds for all $\ell \in \mathbb{N}_0$.

(*c*) Let ℓ and m be positive integers. Assume that x and y are arbitrary points in $\mathbb{E}^n$ such that $\|x - y\| = \frac{\ell}{m}\rho$. Then, we assert that $\|f(x) - f(y)\| = \frac{\ell}{m}\rho$. Since $n > 1$ and $\|x - y\| < 2\ell\rho$, there exists a point z in $\mathbb{E}^n$ such that $\|z - x\| = \ell\rho = \|z - y\|$. Hence, we can choose the points a and b of $\mathbb{E}^n$ as follows

$$x = z + \ell(a - z) \quad \text{and} \quad y = z + \ell(b - z), \tag{2.17}$$

and we set

$$x' = z + m(a - z) \quad \text{and} \quad y' = z + m(b - z). \tag{2.18}$$

Since $\|a - z\| = \rho = \|b - z\|$, it follows from (*b*), (2.17), and (2.18) that

$$\begin{aligned} f(x) &= f\big(z + \ell(a - z)\big) = f(z) + \ell\big(f(a) - f(z)\big), \\ f(y) &= f\big(z + \ell(b - z)\big) = f(z) + \ell\big(f(b) - f(z)\big), \\ f(x') &= f\big(z + m(a - z)\big) = f(z) + m\big(f(a) - f(z)\big), \\ f(y') &= f\big(z + m(b - z)\big) = f(z) + m\big(f(b) - f(z)\big). \end{aligned} \tag{2.19}$$

Since $\|a - b\| = \frac{1}{\ell}\|x - y\| = \frac{1}{m}\rho$, we have $\|x' - y'\| = m\|a - b\| = \rho$. By (*a*), we obtain $\|f(x') - f(y')\| = \rho$. According to the first two equalities of (2.19), we get

$$\|f(x) - f(y)\| = \ell \|f(a) - f(b)\|.$$

Due to the last two equalities of (2.19), we obtain

$$\rho = \|f(x') - f(y')\| = m\|f(a) - f(b)\|.$$

The last two equalities imply $\|f(x) - f(y)\| = \frac{\ell}{m}\rho$.

(d) Let x and y be arbitrary points in $\mathbb{E}^n$ such that $r\rho < \|x - y\| < s\rho$, where r and s are positive rational numbers. We will prove that $r\rho \leq \|f(x) - f(y)\| \leq s\rho$. Since $n > 1$ and $\|x - y\| < s\rho$, there is a point z with $\|z - x\| = \frac{s}{2}\rho = \|z - y\|$. It follows from (c) that $\|f(z) - f(x)\| = \frac{s}{2}\rho = \|f(z) - f(y)\|$, and hence,

$$\|f(x) - f(y)\| \leq \|f(x) - f(z)\| + \|f(z) - f(y)\| = s\rho.$$

We set $w = x + \frac{s\rho}{\|y-x\|}(y - x)$. Then, we see that $\|w - x\| = s\rho$ and

$$\|w - y\| = \left(\frac{s\rho}{\|y - x\|} - 1\right)\|y - x\| = s\rho - \|y - x\| < (s - r)\rho.$$

Hence, it follows from (c) and the first part of (d) that $\|f(w) - f(x)\| = s\rho$ and $\|f(w) - f(y)\| \leq (s - r)\rho$, which implies that

$$\|f(x) - f(y)\| \geq \|f(x) - f(w)\| - \|f(y) - f(w)\| \geq s\rho - (s - r)\rho = r\rho.$$

(e) Let x and y be distinct points of $\mathbb{E}^n$. We note that there are some monotonic sequences $\{r_i\}$ and $\{s_i\}$ consisting of positive rational numbers with the following properties:

- $r_i\rho < \|x - y\| < s_i\rho$ for all $i \in \mathbb{N}$;
- $\lim_{i\to\infty} r_i\rho = \|x - y\|$;
- $\lim_{i\to\infty} s_i\rho = \|x - y\|$.

Therefore, it follows from (d) that

$$\|x - y\| = \lim_{i\to\infty} r_i\rho \leq \|f(x) - f(y)\| \leq \lim_{i\to\infty} s_i\rho = \|x - y\|,$$

which implies that f is an isometry. Finally, due to the theorem of Baker, f is affine.

□

We recall that

$$\gamma(\alpha, \beta) = 4\alpha^2 - 2\beta^2\left(1 - \frac{1}{n}\right) > 0$$

for some appropriate positive real numbers α and β (see Lemma 2.21). In the following lemma, we set $\varepsilon = \sqrt{\gamma(\alpha, \beta)}$.

Lemma 2.24 *Let n be a fixed integer larger than 1. Assume that α and β are positive real numbers that satisfy $\gamma(\alpha, \beta) > 0$. Moreover, assume that a mapping $f : \mathbb{E}^n \to \mathbb{E}^n$ preserves the distances α and β and that x and y are points of $\mathbb{E}^n$ with $\|x - y\| = \varepsilon$, where $\varepsilon = \sqrt{\gamma(\alpha, \beta)}$. Then, $\|f(x) - f(y)\| \in \{0, \varepsilon\}$. In particular, if $2\varepsilon > \alpha$, then $\|f(x) - f(y)\| = \varepsilon$.*

Proof Without loss of generality, we assume that $\varepsilon \neq \alpha$. Let x and y be the α-associated points of a β-set P. Since f preserves the distance β, $P' = f(P)$ is also a β-set. If we denote the α-associated points of P' by x' and y', then $f(x), f(y) \in \{x', y'\}$ because f preserves the distance α and the α-associated points of P' are uniquely determined. According to Lemma 2.21 and the previous sentence, we have $\|f(x) - f(y)\| \in \{0, \|x' - y'\|\} = \{0, \varepsilon\}$.

Assume that $2\varepsilon > \alpha$. We assert that $f(x) \neq f(y)$. On the contrary, suppose $f(x) = f(y)$ and select a z in $\mathbb{E}^n$ with $\|z - x\| = \varepsilon$ and $\|y - z\| = \alpha$. It follows from the first part of this proof that $\|f(x) - f(z)\| \in \{0, \varepsilon\}$. That is, $\|f(y) - f(z)\| \in \{0, \varepsilon\}$ because of $f(x) = f(y)$. Thus, $\alpha = \|y - z\| = \|f(y) - f(z)\| \in \{0, \varepsilon\}$, which is contrary to our assumption that $\varepsilon \neq \alpha$. □

Finally, we have all the tools to prove the Beckman–Quarles theorem.

Theorem 2.17 (Beckman and Quarles) *Let n be a fixed integer greater than 1. If a mapping $f : \mathbb{E}^n \to \mathbb{E}^n$ preserves a distance $\rho > 0$, then f is an affine isometry.*

Proof

(*a*) We note that $\gamma(\rho, \rho) = 2(1 + \frac{1}{n})\rho^2 > 0$, $\varepsilon := \sqrt{\gamma(\rho, \rho)} = \sqrt{2(1 + \frac{1}{n})}\rho$, and that $2\varepsilon > \rho$. If we put $\alpha = \beta = \rho$ in Lemma 2.24, then we see that f preserves the distance $\sqrt{2(1 + \frac{1}{n})}\rho$.

(*b*) We note that

$$\gamma\left(\sqrt{2\left(1 + \frac{1}{n}\right)}\rho, \sqrt{2\left(1 + \frac{1}{n}\right)}\rho\right) = 4\left(1 + \frac{1}{n}\right)^2 \rho^2 > 0,$$

$$\varepsilon := \gamma\left(\sqrt{2\left(1 + \frac{1}{n}\right)}\rho, \sqrt{2\left(1 + \frac{1}{n}\right)}\rho\right)^{1/2} = 2\left(1 + \frac{1}{n}\right)\rho,$$

and that $2\varepsilon > \rho$. If we put $\alpha = \beta = \sqrt{2\left(1 + \frac{1}{n}\right)}\rho$ in Lemma 2.24, then we can check that f preserves the distance $2\left(1 + \frac{1}{n}\right)\rho$.

(c) Now, if we set $\alpha = \rho$ and $\beta = \sqrt{2\left(1 + \frac{1}{n}\right)}\rho$ in Lemma 2.24, then we can see that $\|f(x) - f(y)\| \in \{0, \frac{2}{n}\rho\}$ whenever the distance between x and y is $\frac{2}{n}\rho$. That is, $\|f(x) - f(y)\| \leq \frac{2}{n}\rho$ whenever the distance between x and y is $\frac{2}{n}\rho$.

(d) Finally, if we replace ρ with $\frac{2}{n}\rho$ and we set $N = n + 1$ in Proposition 2.23, then we may check that f is an affine isometry. □

We must now mention some historically important references. Independently of the aforementioned proof of the Beckman-Quarles theorem, in 1970, C. G. Townsend [66] proved the weakened Beckman–Quarles theorem for $n = 2$:

If an injective mapping $f : \mathbb{E}^2 \to \mathbb{E}^2$ preserves the unit distance, then f is an affine isometry.

R. L. Bishop, who did not satisfy with some of the unproven claims in Townsend's paper, proved the Beckman–Quarles theorem on his own in 1973 (see [9]).

Finally, it may be worth mentioning that V. Totik [65] recently presented a short and elementary proof of the Beckman–Quarles theorem by using only the triangle inequality but no calculation.

3 Aleksandrov-Benz Problem

Abstract

In 1970, A. D. Aleksandrov asked whether a mapping must be an isometry if it preserves a certain distance. As we saw in the previous chapter, F. S. Beckman and D. A. Quarles solved this problem for mappings from an n-dimensional Euclidean space into the same one. After weakening the result of the Beckman–Quarles theorem to make the proof easier, E. M. Schröder proved in 1979 that any mapping that preserves two distances ρ and 2ρ is an affine isometry. In a situation where the Beckman–Quarles theorem is already known, Schröder's theorem by itself is of little significance. Schröder, however, presented a novel idea like the m-chain in the process of proving his theorem, and using this idea, W. Benz was able to significantly expand the Beckman–Quarles theorem. In this chapter, we present in detail three historically important theorems, Schröder's theorem, Benz's theorem, and Benz–Berens theorem.

3.1 Theorem of Schröder

After about a quarter of a century, E. M. Schröder [61] supplemented the Beckman–Quarles theorem by showing that any mapping that preserves two distances is an isometry, and he suggested an easier way to prove the theorem of Beckman and Quarles. We will introduce Schröder's theorem here.

Theorem 3.1 (Schröder) *Let n be a fixed integer greater than 1, and let ρ be a fixed positive real number. If a mapping $f : \mathbb{E}^n \to \mathbb{E}^n$ preserves two distances ρ and 2ρ, then f is an affine isometry.*

S.-M. Jung, *Aleksandrov-Rassias Problems on Distance Preserving Mappings*, Frontiers in Mathematics, https://doi.org/10.1007/978-3-031-77613-7_3

Proof

(*a*) For each $m \in \mathbb{N}$, let $(x_0, x_1, \ldots, x_m)$ be an $(m+1)$-tuple of $m+1$ distinct collinear points in $\mathbb{E}^n$ with $\|x_{i-1} - x_i\| = \rho$ for all $i \in \{1, 2, \ldots, m\}$. The tuple $(x_0, x_1, \ldots, x_m)$ is called an m-chain.

We assert that f maps each m-chain onto an m-chain. Let $(x_0, x_1, \ldots, x_m)$ be an arbitrary m-chain, where $m > 1$. Then, we have

$$\|x_i - x_{i+1}\| = \|x_{i+1} - x_{i+2}\| = \rho$$

for all $i \in \{0, 1, \ldots, m-2\}$. Since $(x_0, x_1, \ldots, x_m)$ is an m-chain, if we set $a = x_i$, $b = x_{i+2}$, and $m = x_{i+1}$ in Lemma 2.20, then it follows that $x_{i+1} = \frac{1}{2}(x_i + x_{i+2})$ for each $i \in \{0, 1, \ldots, m-2\}$ and

$$\|x_i - x_{i+2}\| = 2\rho$$

for any $i \in \{0, 1, \ldots, m-2\}$. Since ρ and 2ρ are preserved by f, we obtain

$$\|f(x_i) - f(x_{i+1})\| = \|f(x_{i+1}) - f(x_{i+2})\| = \rho \quad \text{and} \quad \|f(x_i) - f(x_{i+2})\| = 2\rho,$$

and hence, by the strict convexity of $\mathbb{E}^n$, we have

$$f(x_{i+1}) = \frac{1}{2}\big(f(x_i) + f(x_{i+2})\big)$$

for all $i \in \{0, 1, \ldots, m-2\}$. Therefore, we see that $(f(x_i), f(x_{i+1}), f(x_{i+2}))$ is a 2-chain for each $i \in \{0, 1, \ldots, m-2\}$, which means that $(f(x_0), f(x_1), \ldots, f(x_m))$ is an m-chain.

(*b*) We will now prove that f preserves the distance $r\rho$ for each positive rational number r. Let $r = \frac{\ell}{m}$ be an arbitrary rational number, where $\ell, m \in \mathbb{N}$. We assume that arbitrary points x and y in $\mathbb{E}^n$ satisfy the condition $\|x - y\| = \frac{\ell}{m}\rho = r\rho$. Then there exists a point z in $\mathbb{E}^n$ such that $\|x - z\| = \|y - z\| = \ell\rho$. We set $x' = z + \frac{m}{\ell}(x - z)$ and $y' = z + \frac{m}{\ell}(y - z)$. Then, the points z, x, x' resp. z, y, y' are collinear and belong to an ℓm-chain $(x_0, \ldots, x_{\ell m})$ resp. $(y_0, \ldots, y_{\ell m})$, where $x_0 = y_0 = z$, $x_\ell = x$, $x_m = x'$, $y_\ell = y$, and $y_m = y'$. Since $\|x' - y'\| = \rho$, it follows from (*a*) that $\|f(x) - f(y)\| = \frac{\ell}{m}\rho = r\rho$.

(*c*) For any $r > 0$ and $x \in \mathbb{E}^n$, we denote by $\overline{B}_r(x)$ the closed ball in $\mathbb{E}^n$ of radius r and centered at x:

$$\overline{B}_r(x) = \big\{z \in \mathbb{E}^n : \|z - x\| \leq r\big\}.$$

We claim that $f(\overline{B}_{r\rho}(x)) \subset \overline{B}_{r\rho}(f(x))$ for all $x \in \mathbb{E}^n$ and for all rational numbers $r > 0$. For each $z \in \overline{B}_{r\rho}(x)$, there is a point $y \in \mathbb{E}^n$ such that $\|y-x\| = \|z-y\| = \frac{r}{2}\rho$. Hence, it follows from (b) that $\|f(y) - f(x)\| = \|f(z) - f(y)\| = \frac{r}{2}\rho$. Thus, using the triangle inequality, we have

$$\|f(z) - f(x)\| \leq \|f(z) - f(y)\| + \|f(y) - f(x)\| = r\rho,$$

i.e., $f(z) \in \overline{B}_{r\rho}(f(x))$, which implies that $f(\overline{B}_{r\rho}(x)) \subset \overline{B}_{r\rho}(f(x))$.

(d) Finally, we assert that $\|f(x) - f(y)\| = \|x - y\|$ for all $x, y \in \mathbb{E}^n$. If there would exist some $x, y \in \mathbb{E}^n$ such that $\|f(x) - f(y)\| > \|x - y\|$, then there would exist some positive rational number r such that $\|x - y\| \leq r\rho < \|f(x) - f(y)\|$. Then, it would follow that $f(y) \notin \overline{B}_{r\rho}(f(x))$, which would be contrary to (c). On the other hand, we assume that there would exist some $x, y \in \mathbb{E}^n$ such that $\|f(x) - f(y)\| < \|x - y\|$. Let z be a point on the line going through x and y such that $\|x - z\| = r\rho$ for some rational number $r > 0$ and such that y lies between x and z. Then, it would be true that $\|x - y\| < r\rho$. Thus, there would exist a rational number $s > 0$ such that

$$r\rho - \|x - y\| \leq s\rho < r\rho - \|f(x) - f(y)\|.$$

Since $\|y - z\| = r\rho - \|x - y\| \leq s\rho$, we would obtain $y \in \overline{B}_{s\rho}(z)$. Now, by (b) and (c),

$$\begin{aligned} r\rho &= \|x - z\| \\ &= \|f(x) - f(z)\| \\ &\leq \|f(x) - f(y)\| + \|f(y) - f(z)\| \\ &\leq \|f(x) - f(y)\| + s\rho, \end{aligned}$$

which would be contrary to $s\rho < r\rho - \|f(x) - f(y)\|$. Therefore, we conclude that $\|f(x) - f(y)\| = \|x - y\|$ holds for all $x, y \in \mathbb{E}^n$, i.e., f is an isometry. Since $\mathbb{E}^n$ is strictly convex, it follows from the Baker's theorem that f is affine. □

Corollary 3.2 *Let n be a fixed integer satisfying $n > 1$, let ρ be a fixed positive real number, and let $f : \mathbb{E}^n \to \mathbb{E}^n$ be a mapping that has the following property*

$$\|x - y\| = \rho \quad \textit{if and only if} \quad \|f(x) - f(y)\| = \rho \tag{3.1}$$

for any pair of points $x, y \in \mathbb{E}^n$. Then, f is an affine isometry.

Proof For any pair of points $x, y \in \mathbb{E}^n$ with $\|x - y\| = 2\rho$, there exists a unique point z in $\mathbb{E}^n$ such that $\|z - x\| = \|z - y\| = \rho$. If $\|f(x) - f(y)\| \neq 2\rho$, then $f(x)$, $f(y)$ and

$f(z)$ would form the vertices of some (non-degenerate) triangle. Since $n > 1$, there would exist a point w in $\mathbb{E}^n$ such that $f(w) \neq f(z)$ and $\|f(w) - f(x)\| = \|f(w) - f(y)\| = \rho$. It would then follow from (3.1) that $\|w - x\| = \|w - y\| = \rho$, i.e., $w = z$ and $f(w) = f(z)$, which would contradict the choice of w. Hence, f preserves the distances ρ and 2ρ by (3.1) and the argument aforementioned. Therefore, it follows from Theorem 3.1 that f is an affine isometry. □

Instead of literally saying the condition (3.1), we will say that f preserves the distance ρ *in both directions*.

3.2 Theorem of Benz

E. M. Schröder shows that in the case of a Euclidean space $X = Y = \mathbb{E}^n$ with $n > 1$, a mapping $f : \mathbb{E}^n \to \mathbb{E}^n$ is an affine isometry if and only if f preserves two distances ρ and 2ρ, where $\rho > 0$ is a real number.

In the proof of the following theorem, which significantly extends Schröder's theorem, the idea of m-chains developed by Schröder is widely used.

Theorem 3.3 (Benz) *Let $\rho > 0$ and $N > 1$ be a fixed real number and an integer, respectively. Assume that X and Y are real normed spaces with the following properties:*

- *(i) For any $a \in X$ with $\|a\| < 1$, there is $b \in X$ with $\|a - b\| = 1 = \|a + b\|$;*
- *(ii) If $c, d \in Y$ satisfy $\|c\| = 1 = \|d\|$ and $\|c + d\| = 2$, then $c = d$.*

If the distance ρ is contractive and the distance $N\rho$ is extensive by a mapping $f : X \to Y$, then f is an affine isometry.

Proof We carry out the proof of this theorem in several steps.

(a) Suppose x and y are arbitrary elements of X such that $\|x - y\| = \rho$. We claim that $\|f(x) - f(y)\| = \rho$. If we set $p_i = y + i(x - y)$ for each $i \in \{0, 1, \ldots, N\}$, then we have $\|p_N - y\| = N\rho$ and $\|p_i - p_{i-1}\| = \rho$ for all $i \in \{1, 2, \ldots, N\}$. That is, $(p_0, p_1, \ldots, p_N)$ is an N-chain. Since $N\rho$ is extensive and ρ is contractive by f, it holds that $\|f(p_N) - f(y)\| \geq N\rho$ and $\|f(p_i) - f(p_{i-1})\| \leq \rho$ for each $i \in \{1, 2, \ldots, N\}$. By using these facts, together with the triangle inequality, we obtain

$$N\rho \leq \|f(p_N) - f(y)\| \leq \sum_{i=1}^{N} \|f(p_{N+1-i}) - f(p_{N-i})\| \leq N\rho.$$

Thus, we get $\|f(x) - f(y)\| = \|f(p_1) - f(p_0)\| = \rho$.

(*b*) Suppose x and y are arbitrary elements of X such that $\|x - y\| = 2\rho$. We claim that $\|f(x) - f(y)\| = 2\rho$. If we set $p_i = y + \frac{i}{2}(x - y)$ for all $i \in \{0, 1, \ldots, N\}$, then we obtain $\|p_N - y\| = N\rho$ and $\|p_i - p_{i-1}\| = \rho$ for any $i \in \{1, 2, \ldots, N\}$. Indeed, $(p_0, p_1, \ldots, p_N)$ is an N-chain. Using the triangle inequality and applying (*a*), we have

$$N\rho \leq \|f(p_N) - f(y)\| \leq \sum_{i=1}^{N} \|f(p_{N+1-i}) - f(p_{N-i})\| = N\rho,$$

i.e.,

$$\|f(p_N) - f(y)\| = \sum_{i=1}^{N} \|f(p_{N+1-i}) - f(p_{N-i})\|. \tag{3.2}$$

We will now show that

$$\begin{aligned} \|f(x) - f(y)\| &= \|f(p_2) - f(p_0)\| \\ &= \|f(p_2) - f(p_1)\| + \|f(p_1) - f(p_0)\|. \end{aligned} \tag{3.3}$$

If the left side of (3.3) were smaller than the right side, we would have $N > 2$ in (3.2) and due to the triangle inequality

$$\begin{aligned} \|f(p_N) - f(y)\| &\leq \sum_{i=1}^{N-2} \|f(p_{N+1-i}) - f(p_{N-i})\| + \|f(p_2) - f(p_0)\| \\ &< \sum_{i=1}^{N} \|f(p_{N+1-i}) - f(p_{N-i})\|, \end{aligned}$$

which contradicts (3.2). Hence, the equality holds in (3.3), i.e., we have

$$\|f(x) - f(y)\| = \|f(p_2) - f(p_1)\| + \|f(p_1) - f(p_0)\| = 2\rho.$$

(*c*) Suppose a, b, c are any elements of Y such that

$$\|b - a\| = \rho = \|c - b\| \quad \text{and} \quad \|c - a\| = 2\rho.$$

We claim that $c = 2b - a$. Since

$$\left\|\frac{1}{\rho}(b - a)\right\| = 1 = \left\|\frac{1}{\rho}(c - b)\right\| \quad \text{and} \quad \left\|\frac{1}{\rho}(c - a)\right\| = 2,$$

it follows from (ii) that $\frac{1}{\rho}(b-a)=\frac{1}{\rho}(c-b)$, i.e., $c=2b-a$.

(d) Suppose x and y are arbitrary elements of X such that $\|x-y\|=\rho$. We claim that $f(x+m(y-x))=f(x)+m(f(y)-f(x))$ for all $m\in\mathbb{N}_0$. Our assertion obviously holds for $m\in\{0,1\}$. Suppose our claim holds for $m>0$ and set $p_m=x+m(y-x)$. Since

$$\|p_m-p_{m-1}\|=\rho=\|p_{m+1}-p_m\| \quad\text{and}\quad \|p_{m+1}-p_{m-1}\|=2\rho,$$

it follows from (a) and (b) that

$$\|f(p_m)-f(p_{m-1})\|=\rho=\|f(p_{m+1})-f(p_m)\|$$

and

$$\|f(p_{m+1})-f(p_{m-1})\|=2\rho.$$

But it follows from (c) that

$$f(p_{m+1})=2f(p_m)-f(p_{m-1}).$$

Therefore, we have

$$\begin{aligned}
&f\big(x+(m+1)(y-x)\big)\\
&\quad=2f\big(x+m(y-x)\big)-f\big(x+(m-1)(y-x)\big)\\
&\quad=2f(x)+2m\big(f(y)-f(x)\big)-f(x)-(m-1)\big(f(y)-f(x)\big)\\
&\quad=f(x)+(m+1)\big(f(y)-f(x)\big),
\end{aligned}$$

which implies that our claim also applies to $m+1$. Therefore, by mathematical induction, we conclude that our claim is true.

(e) Suppose α and β are arbitrary positive real numbers such that $2\beta\geq\alpha$. Furthermore, suppose x and y are arbitrary elements of X such that $\|x-y\|=\alpha$. We claim that there is a $z\in X$, which satisfies $\|z-x\|=\beta=\|z-y\|$. If $2\beta=\alpha$ we prove our claim by choosing $z=\frac{1}{2}(x+y)$.

Now suppose that $2\beta>\alpha$. In this case, we set $a=\frac{1}{2\beta}(y-x)$. Then we have $\|a\|<1$. Due to (i), there is also a $b\in X$ such that $\|a-b\|=1=\|a+b\|$. If we set $z=\frac{1}{2}(x+y)+\beta b$, then $\|z-x\|=\beta\|a+b\|=\beta$ and $\|z-y\|=\beta\|-a+b\|=\beta$.

By applying the proof ideas of Theorem 3.1 to the case at hand, we now come to the following conclusion:

(f) Suppose $m,n\in\mathbb{N}$ and $x,y\in X$ satisfy $\|x-y\|=\frac{m}{n}\rho$. We claim that $\|f(x)-f(y)\|=\frac{m}{n}\rho$. If we set $\alpha=\frac{m}{n}\rho$ and $\beta=m\rho$, then (e) implies the existence of a $z\in X$ with $\|x-z\|=m\rho=\|y-z\|$. Choose $a,b\in X$ such that

$$x = z + m(a - z), \quad y = z + m(b - z) \tag{3.4}$$

and we set

$$x' = z + n(a - z), \quad y' = z + n(b - z). \tag{3.5}$$

Then, by (3.4) and (3.5), we have $\|a - z\| = \|b - z\| = \|x' - y'\| = \rho$.

Moreover, using (d), we obtain

$$f(x) = f(z) + m\big(f(a) - f(z)\big), \quad f(y) = f(z) + m\big(f(b) - f(z)\big) \tag{3.6}$$

and

$$f(x') = f(z) + n\big(f(a) - f(z)\big), \quad f(y') = f(z) + n\big(f(b) - f(z)\big). \tag{3.7}$$

Hence, it follows from (a) and (3.7) that

$$\|x' - y'\| = \rho = \|f(x') - f(y')\| = n\|f(a) - f(b)\|.$$

In addition, it follows from (3.6) that

$$\|f(x) - f(y)\| = m\|f(a) - f(b)\|,$$

i.e., $\|f(x) - f(y)\| = \frac{m}{n}\rho$.

(g) Suppose r_1 and r_2 are rational numbers with $0 < r_1 < r_2$. We claim that $r_1\rho \leq \|f(x) - f(y)\| \leq r_2\rho$ for any $x, y \in X$ with $r_1\rho < \|x - y\| < r_2\rho$.

$(g.1)$ Suppose x and y are arbitrary elements of X with $r_1\rho < \|x - y\| < r_2\rho$. Due to (e) for $\alpha = r_1\rho$ and $\beta = \frac{r_2}{2}\rho$, there exists a $z \in X$ such that

$$\|x - z\| = \frac{r_2}{2}\rho = \|y - z\|.$$

It then follows from (f) that

$$\|f(x) - f(z)\| = \frac{r_2}{2}\rho = \|f(y) - f(z)\|.$$

Thus, we obtain

$$\|f(x) - f(y)\| \leq \|f(x) - f(z)\| + \|f(z) - f(y)\| = r_2\rho.$$

$(g.2)$ Suppose there were $x, y \in X$ such that $r_1\rho < \|x - y\| < r_2\rho$ and $\|f(x) - f(y)\| < r_1\rho$. Then we have

$$r_2\rho - \|x - y\| < (r_2 - r_1)\rho < r_2\rho - \|f(x) - f(y)\|. \tag{3.8}$$

We set $z = x + \lambda(y - x)$, where $\lambda = \frac{r_2\rho}{\|x-y\|} > 1$. Then, we obtain $\|z - x\| = r_2\rho$, and it follows from (3.8) that $\|z - y\| = (\lambda - 1)\|x - y\| < (r_2 - r_1)\rho$. Hence, using (f), we have $\|f(z) - f(x)\| = r_2\rho$. It would now follow from $(g.1)$ that

$$r_2\rho = \|f(z) - f(x)\| \le \|f(z) - f(y)\| + \|f(y) - f(x)\| < (r_2 - r_1)\rho + r_1\rho,$$

a contradiction. Therefore, we can conclude that $\|f(x) - f(y)\| \ge r_1\rho$.

(h) Assume that $\{\alpha_i\}$ is a strictly increasing sequence and $\{\beta_i\}$ is a strictly decreasing sequence with the following properties:
(iii) $\alpha_i, \beta_i \in \mathbb{Q}$ for all $i \in \mathbb{N}$;
(iv) $\lim_{i\to\infty} \alpha_i = 1 = \lim_{i\to\infty} \beta_i$.
It then follows from (g) that, for every $i \in \mathbb{N}$, if $\alpha_i\rho < \|x - y\| < \beta_i\rho$, then $\alpha_i\rho \le \|f(x) - f(y)\| \le \beta_i\rho$, which implies that f is an isometry from X into Y.

Suppose x and y are nonzero elements of Y with $\|x+y\| = \|x\|+\|y\|$. According to Lemma 2.8, we can set $s = \frac{1}{\|x\|}$ and $t = \frac{1}{\|y\|}$ such that $\|sx + ty\| = \|sx\| + \|ty\|$. Then, our hypothesis (ii) for $c = sx$ and $d = ty$ implies $sx = ty$, i.e., $x = \frac{t}{s}y$ with $\frac{t}{s} > 0$. Due to Definition 2.4, Y is strictly convex. Finally, on account of the theorem of Baker, f is affine. □

Remark 3.4 Every real inner product space has the properties (i) and (ii) given in Theorem 3.3 if its dimension is greater than 1.

Proof Let $(X, \langle\cdot,\cdot\rangle)$ be a real inner product space with $\dim X > 1$. Assume that a is an arbitrary element of X with $\|a\| < 1$. Since $\dim X > 1$, there is an element b orthogonal to a such that $\|b\|^2 = 1 - \|a\|^2$. Then we have

$$\|a - b\|^2 = \langle a - b, a - b\rangle = \|a\|^2 - 2\langle a, b\rangle + \|b\|^2 = 1$$

and

$$\|a + b\|^2 = \langle a + b, a + b\rangle = \|a\|^2 + 2\langle a, b\rangle + \|b\|^2 = 1.$$

That is, $(X, \langle\cdot,\cdot\rangle)$ has the property (i).

Let $(Y, \langle\cdot,\cdot\rangle)$ be a real inner product space with $\dim Y > 1$. Suppose c and d are arbitrary elements of Y such that $\|c\| = 1 = \|d\|$ and $\|c + d\| = 2$. On the contrary, assume that $c \ne d$. Then we obtain

$$4 = \|c + d\|^2 = \langle c + d, c + d\rangle = \|c\|^2 + 2\langle c, d\rangle + \|d\|^2 = 2 + 2\langle c, d\rangle.$$

Hence, $\langle c, d\rangle = 1$. According to the Cauchy-Schwarz inequality, the vectors c and d are linearly dependent. Thus, there is a real number $\alpha \neq 0$ such that $c = \alpha d$. Since $\|c\| = 1 = \|d\|$ and $c \neq d$, it follows that $\alpha = -1$. Moreover, we have $\|c + d\| = \|0\| = 0 \neq 2$, a contradiction. Hence, c and d must be the same. Therefore, $(Y, \langle\cdot,\cdot\rangle)$ has the property (ii). □

According to Remark 3.4, the Euclidean space $\mathbb{E}^n$ for $n > 1$ has the properties (i) and (ii), which are given in the theorem of Benz. It follows from (a) of the proof of Beckman–Quarles theorem that if f preserves $\rho > 0$, then it also preserves $\sqrt{2(1+\frac{1}{n})}\rho$. Due to (b) of the proof of Beckman–Quarles theorem, f preserves $2\left(1+\frac{1}{n}\right)\rho$. Moreover, by (c) of the proof, the distance $\frac{2}{n}\rho$ is contractive by f. In summary, $\frac{2}{n}\rho$ is contractive and $2\left(1+\frac{1}{n}\right)\rho$ is extensive by f. If we replace ρ and N with $\frac{2}{n}\rho$ and $n+1$, respectively, Benz's theorem provides the proof of the Beckman–Quarles theorem. This fact implies that Benz's theorem is a generalization of the Beckman–Quarles theorem.

3.3 Theorem of Benz and Berens

In the last part of the proof of Theorem 3.3, we showed that the second condition of the theorem implies the strict convexity of the range space Y.

If the range space is a real normed space that is strictly convex, W. Benz and H. Berens [7] could generalize both Schröder's and Benz's theorems by weakening their conditions.

The proof of the following theorem is similar to the proof of Benz's theorem. We will omit the parts that are similar to each other.

Theorem 3.5 (Benz and Berens) *Let X be a real normed space with $\dim X > 1$ and let Y be a real normed space, which is strictly convex. Suppose $f : X \to Y$ is a mapping and $N > 1$ is a fixed integer. If a distance $\rho > 0$ is contractive and $N\rho$ is extensive by f, then f is an affine isometry.*

Proof We divide the proof of this theorem into several steps and prove each step one by one.

(a) Following the part (a) of the proof of Theorem 3.3, we can prove that $\|f(x) - f(y)\| = \rho$ for any vectors x and y in X with $\|x - y\| = \rho$.

(b) Now, we assume that x and y are arbitrary vectors in X with $\|x - y\| = 2\rho$. It follows from the part (b) of the proof of Theorem 3.3 that $\|f(x) - f(y)\| = 2\rho$.

(c) We assume that x, y, and z are arbitrary vectors in Y such that $\|y - x\| = \rho = \|z - y\|$ and $\|z - x\| = 2\rho$. Since Y is strictly convex, it follows from the facts

$$\|(z - y) + (y - x)\| = \|z - x\| = 2\rho = \|z - y\| + \|y - x\|$$

and

$$\|z - y\| = \|y - x\|$$

that $z - y = y - x$, i.e., $z = 2y - x$.

(d) Suppose x and y are arbitrary vectors in X with $\|x - y\| = \rho$. Just as with the part (d) of the proof of Theorem 3.3, we use mathematical induction to prove

$$f\big(x + m(y - x)\big) = f(x) + m\big(f(y) - f(x)\big)$$

for all $m \in \mathbb{N}_0$.

(e) Let α and β be positive real numbers with $2\beta \geq \alpha$. Furthermore, assume that x and y are arbitrary vectors in X with $\|x - y\| = \alpha$. We claim that there exists a $z \in X$ with $\|z - x\| = \beta = \|z - y\|$.

If $2\beta = \alpha$, then we just set $z = \frac{1}{2}(x + y)$. Otherwise, if we set $a = \frac{1}{2\beta}(y - x)$, then $\|a\| < 1$. Since X is a real normed space with $\dim X > 1$, there exists a $b \in X$ with $\|a - b\| = 1 = \|a + b\|$. We now put $z = \frac{1}{2}(x + y) + \beta b$. Then, we have $\|z - x\| = \beta\|a + b\| = \beta$ and $\|z - y\| = \beta\|b - a\| = \beta$.

(f) Assume that x and y are arbitrary vectors in X and m and $n > 1$ are positive integers such that $\|x - y\| = \frac{m}{n}\rho$. We prove that $\|f(x) - f(y)\| = \frac{m}{n}\rho$.

Due to (e), there exists a $z \in X$ with $\|z - x\| = m\rho = \|z - y\|$. We choose $a, b \in X$ such that

$$x = z + m(a - z) \quad \text{and} \quad y = z + m(b - z) \tag{3.9}$$

and we put

$$x' = z + n(a - z) \quad \text{and} \quad y' = z + n(b - z). \tag{3.10}$$

Then we have $\|a - z\| = \|b - z\| = \|x' - y'\| = \rho$.

It follows from (d) that

$$\begin{cases} f(x) = f(z) + m\big(f(a) - f(z)\big), \\ f(y) = f(z) + m\big(f(b) - f(z)\big) \end{cases} \tag{3.11}$$

and

$$\begin{cases} f(x') = f(z) + n\big(f(a) - f(z)\big), \\ f(y') = f(z) + n\big(f(b) - f(z)\big). \end{cases} \tag{3.12}$$

Moreover, by (3.9), we get $m\|a-b\| = \|x-y\| = \frac{m}{n}\rho$, and thus, $\|a-b\| = \frac{1}{n}\rho$. In view of (3.10), we obtain $\|x'-y'\| = n\|a-b\| = \rho$. Thus, due to (a), we get $\|f(x')-f(y')\| = \rho$. On the other hand, using (3.12), we obtain $n\|f(a)-f(b)\| = \|f(x')-f(y')\| = \rho$, and thus, $\|f(a)-f(b)\| = \frac{1}{n}\rho$. Furthermore, using (3.11), we obtain $\|f(x)-f(y)\| = m\|f(a)-f(b)\| = \frac{m}{n}\rho$.

(g) Let x and y be arbitrary vectors in X and let r_1 and r_2 be positive rational numbers with $r_1\rho < \|x-y\| < r_2\rho$. By the same way as the part (g) of the proof of Theorem 3.3, we can verify that $r_1\rho \leq \|f(x)-f(y)\| \leq r_2\rho$.

(h) Finally, by letting $r_1 \uparrow r_2$ in (g), we may conclude that $f : X \to Y$ is an isometry. The affinity property of f is an immediate consequence of Baker's theorem. □

Using the triangle inequality, we can easily show that the conditions given in the theorem of Benz and Berens are equivalent to the condition that f preserves the two distances ρ and $N\rho$.

Considering the three main theorems of this chapter, we can now define the Aleksandrov–Benz problem as follows:

Problem 3.6 (Aleksandrov–Benz Problem) Let X and Y be normed spaces and let $f : X \to Y$ be a mapping. If there exist a real number $\rho > 0$ and an integer $N > 1$ such that f preserves the two distances ρ and $N\rho$, is f then necessarily an isometry?

We remark that the distance ratio in the Aleksandrov–Benz problem is restricted to an integer greater than 1.

Let X and Y be real normed spaces such that $\dim X > 1$ and Y is strictly convex. Also assume that $f : X \to Y$ preserves two distances whose ratio is not an integer. Whether f should be an isometry in this case is still an open question, which we will discuss in detail in the next chapters.

Aleksandrov–Rassias Problems

4

Abstract

If the domain space and the range space are the same Euclidean space with dimension greater than 1, the Aleksandrov problem has already been solved by the Beckman–Quarles theorem. W. Benz and H. Berens also solved the extended Aleksandrov problem under the additional conditions that the domain is a real normed space with dimension greater than 1, the range is a strictly convex real normed space, two distances are preserved, and that the ratio of the two distances is an integer. In this chapter, we investigate the Aleksandrov–Rassias problems, which focus on cases where the domain and range of the mapping involved differ, and cases where the ratio of two distances that are preserved is not an integer. By introducing interesting examples and counterexamples related to these topics, we try to help readers easily grasp the core reality of the problem.

4.1 Aleksandrov–Rassias Problems

As we see in Sect. 3.1, E. M. Schröder proved that if a mapping $f : \mathbb{E}^n \to \mathbb{E}^n$ preserves two distances ρ and 2ρ, then f is an affine isometry. Moreover, the theorem of Schröder was further generalized by W. Benz and H. Berens.

We now introduce the concepts first established by Th. M. Rassias and P. Šemrl in their paper [59].

Definition 4.1 Let X and Y be normed spaces and let $f : X \to Y$ be a mapping.

S.-M. Jung, *Aleksandrov-Rassias Problems on Distance Preserving Mappings*, Frontiers in Mathematics, https://doi.org/10.1007/978-3-031-77613-7_4

(i) We say that f has the *distance one preserving property* if and only if for all $x, y \in X$, $\|x - y\| = 1$ implies $\|f(x) - f(y)\| = 1$. In this case, we usually say that f satisfies (DOPP).

(ii) We say that f has the *strong distance one preserving property* when for all $x, y \in X$, $\|x - y\| = 1$ if and only if $\|f(x) - f(y)\| = 1$. In this case, we usually say that f satisfies (SDOPP).

Th. M. Rassias [57] further generalized the ideas of Schröder and of Benz and Berens by posing the following problems, which are extensions of the Aleksandrov problem:

Problem 4.2 (Aleksandrov–Rassias Problems) Let X and Y be normed spaces.

(i) Is a mapping $f : X \to Y$ that satisfies (DOPP) necessarily an isometry?

(ii) Is a mapping $f : X \to Y$ that preserves two or more distances with non-integer ratios necessarily an isometry?

Such problems are called *Aleksandrov–Rassias problems*. It should be noted that in Problem 4.2 (ii) we constrain the distance ratios to be non-integers. This is because when the distance ratios are integers, it is more appropriate to classify this case as an Aleksandrov–Benz problem rather than an Aleksandrov–Rassias problem.

If $X = Y = \mathbb{R}$, Examples 2.14 and 2.16 show that there are counterexamples to the first Aleksandrov-Rassias problem. Nevertheless, we can ask for a solution to this problem with an additional assumption like the differentiability of f. Unfortunately, the answer is still negative, as shown in an example given in [12].

Example 4.3 Let us define a mapping $f : \mathbb{R} \to \mathbb{R}$ by

$$f(x) = x + \sin(2\pi x).$$

Obviously, f is a differentiable mapping. Moreover, we easily check that

$$\begin{aligned} f(x+1) &= x + 1 + \sin(2\pi x + 2\pi) \\ &= f(x) + 1 \end{aligned}$$

for each $x \in \mathbb{R}$, which implies that f satisfies (DOPP). However, f is not an isometry because, e.g., $f(0) = 0$ and $f(\frac{1}{6}) = \frac{1}{6} + \frac{\sqrt{3}}{2}$.

If $X = Y = \mathbb{E}^n$, the n-dimensional Euclidean space with $1 < n < \infty$, the Beckman–Quarles theorem together with Remark 2.19 gives the positive answer to the first Aleksandrov-Rassias problem. Finally, Example 2.15 gives a negative answer to the first Aleksandrov–Rassias problem for $X = Y = \ell^2$.

4.2 When Domain and Range Are Different

Now we are interested in Problem 4.2 (i) (the first Aleksandrov–Rassias problem) when the domain and the range of the mappings involved are different.

Further, we outline a method to show how to construct examples to prove that for each integer $n > 1$ there are an integer m and a mapping $f : \mathbb{E}^n \to \mathbb{E}^m$ that satisfies (DOPP) but is not an isometry. The following example shows the case of the mapping $f : \mathbb{E}^2 \to \mathbb{E}^8$, which was first introduced in [12] by K. Ciesielski and Th. M. Rassias.

Example 4.4 We can partition the Euclidean plane $\mathbb{E}^2$ into squares of diagonal length 1 as follows:

Now, the nine vertices of the unit regular 8-simplex Σ_8 in $\mathbb{E}^8$ are numbered 1 through 9, and f maps each square denoted by i in Fig. 4.1 to the ith vertex of Σ_8. It is easy to see that when two points lie in the same labeled squares, the distance between these points is different from 1. This mapping $f : \mathbb{E}^2 \to \mathbb{E}^8$ satisfies (DOPP) but is not an isometry.

Using regular hexagons instead of squares, we can construct such a mapping $f : \mathbb{E}^2 \to \mathbb{E}^6$ that satisfies (DOPP) but is not an isometry (see [17]). With this idea, it is easy to construct such examples in higher-dimensional Euclidean spaces.

If a mapping $f : \mathbb{E}^n \to \mathbb{E}^m$ preserves a certain distance, then m is obviously not smaller than n. This is true because $\mathbb{E}^m$ has regular n-simplexes if and only if $m \geq n$. Taking the Beckman–Quarles theorem into account, we now need to examine whether there exists a mapping $f : \mathbb{E}^n \to \mathbb{E}^m$ that satisfies (DOPP) but is not isometric, only when $1 < n < m < \infty$.

Indeed, an interesting theorem concerning this subject was proved by Rassias [56]. Here we present only the outline of his proof.

Theorem 4.5 (Rassias) *For any $n \in \mathbb{N}$, there exists an $m_n \in \mathbb{N}$ such that for each integer $m \geq m_n$, there exists a mapping from $\mathbb{E}^n$ into $\mathbb{E}^m$ which satisfies* (DOPP) *but is not an isometry.*

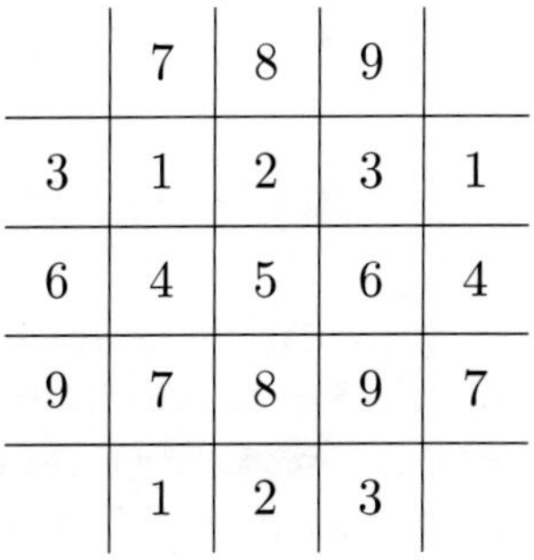

Fig. 4.1 To construct a mapping $f : \mathbb{E}^2 \to \mathbb{E}^8$ that satisfies (DOPP) but is not an isometry, we partition the Euclidean plane as shown in this figure. Each square contains the bottom edge, left edge, and bottom left corner, but none of the other corners

	7	8	9	
3	1	2	3	1
6	4	5	6	4
9	7	8	9	7
	1	2	3	

Proof We partition $\mathbb{E}^n$ into the countably many regions $D_1, D_2, D_3, \ldots$ such that

(i) each region D_i is of diameter smaller than 1;
(ii) any closed n-ball of radius 1 intersects at most k of these regions.

We can choose an integer $m_n > 0$ with the property that the regions $D_1, D_2, D_3, \ldots$ can be partitioned into $m_n + 1$ sets, namely $U_1, U_2, \ldots, U_{m_n+1}$, such that if $x \in D_i$, $y \in D_j$, and D_i and D_j belong to the same U_k, then $d(x, y) \neq 1$.

Let us define a mapping $f : \mathbb{E}^n \to \mathbb{E}^m$ for $m \geq m_n$ such that it maps each set

$$S_k = \bigcup_{D_i \in U_k} D_i$$

to a different vertex of a unit regular m_n-simplex in $\mathbb{E}^m$. It then follows that $d(x, y) = 1$ implies that both x and y are not in the same set S_k. Thus, we have $d(f(x), f(y)) = 1$ for all $x, y \in \mathbb{E}^n$ with $d(x, y) = 1$. Hence, f satisfies (DOPP) but is not an isometry. □

We do not even know if a non-isometric mapping $f : \mathbb{E}^2 \to \mathbb{E}^3$ can satisfy (DOPP). Furthermore, it is still an open problem whether there is a continuous mapping $f : \mathbb{E}^n \to \mathbb{E}^m$, for $m > n$, that satisfies (DOPP) but is not an isometry (see [55, 57]).

Finally, we cannot help but mention the following: In 2003, S.-M. Jung [25] proved the following theorems in connection with the first Aleksandrov–Rassias problem.

Theorem 4.6 *Assume that X and Y are real Hilbert spaces with* $\dim X > 2$ *and* $\dim Y > 2$. *If a mapping* $f : X \to Y$ *preserves a distance* $\rho > 0$ *and if* f *maps the vertices of every square with side lengths* ρ *and* $\sqrt{2}\rho$ *in X onto the vertices of a rhombus with side lengths* ρ *and* $\sqrt{2}\rho$ *in Y, respectively, then* f *is an affine isometry.*

Theorem 4.7 *Let X and Y be real Hilbert spaces with each dimension greater than* 1. *Assume that the distance* $\rho > 0$ *is contractive by a mapping* $f : X \to Y$ *and that there exists an integer* $n > 1$ *such that* $\sqrt{n^2 + 1}\,\rho$ *is extensive by* f. *If* f *maps the midpoint of every line segment joining* v *and* w *of length* $2n\rho$ *into the line segment between* $f(v)$ *and* $f(w)$, *then* f *is an affine isometry.*

4.3 Aleksandrov Problems with Non-standard Metrics

So far we have investigated the Aleksandrov problems mainly in Euclidean spaces. We now consider the Aleksandrov problems by extending them to more general metric spaces, rather than restricting them to the problems in the Euclidean spaces. This section is mainly based on the paper [12] by K. Ciesielski and Th. M. Rassias.

We know that the following classical metrics in $\mathbb{R}^n$ induce the same topology:

$$d_e(x, y) = \left(\sum_{i=1}^{n} (x_i - y_i)^2 \right)^{1/2}, \quad d_m(x, y) = \max_{i \in \{1,2,\ldots,n\}} |x_i - y_i|,$$

and

$$d_\sigma(x, y) = \sum_{i=1}^{n} |x_i - y_i|$$

for all points $x = (x_1, x_2, \ldots, x_n)$ and $y = (y_1, y_2, \ldots, y_n)$ in $\mathbb{R}^n$.

We note that for $n = 1$ all three metrics aforementioned are the same. We now consider the metric space $\mathbb{R}^2$, which has the metric d_m. The following example shows that in this case some mappings satisfy (DOPP) but are not isometries.

Example 4.8 We define a mapping $f : \mathbb{R}^2 \to \mathbb{R}^2$ by $f(x, y) = ([x], [y])$, where $[x]$ denotes the integer part of x. The mapping f maps each point of an appropriate unit square to its lower-left corner such that $f(\mathbb{R}^2) = \mathbb{Z}^2$. Therefore, f satisfies (DOPP) but is not isometric.

Next, we assume that the space $\mathbb{R}^2$ has the metric d_σ. In the following example, we construct a mapping that satisfies (DOPP) but is not isometric.

Example 4.9 We define the mapping $g : (\mathbb{R}^2, d_\sigma) \to (\mathbb{R}^2, d_\sigma)$ by

$$g = \left(\sqrt{2} \cdot R_{\frac{\pi}{4}}\right) \circ f \circ \left(\frac{1}{\sqrt{2}} \cdot R_{\frac{\pi}{4}}^{-1}\right),$$

where f is defined in Example 4.8 and $R_{\pi/4}$ and $R_{\pi/4}^{-1}$ are the rotations defined as

$$R_{\frac{\pi}{4}}(x, y) = \left(\frac{1}{\sqrt{2}}(x + y), \frac{1}{\sqrt{2}}(y - x)\right),$$
$$R_{\frac{\pi}{4}}^{-1}(x, y) = \left(\frac{1}{\sqrt{2}}(x - y), \frac{1}{\sqrt{2}}(x + y)\right).$$

These rotations map unit balls in metric d_m to balls of radius $\sqrt{2}$ with respect to the metric d_σ. Then we have

$$\begin{aligned} g(x, y) &= \left(\sqrt{2} \cdot R_{\frac{\pi}{4}}\right) \circ f \circ \left(\frac{1}{\sqrt{2}} \cdot R_{\frac{\pi}{4}}^{-1}\right)(x, y) \\ &= \left(\left[\frac{x - y}{2}\right] + \left[\frac{x + y}{2}\right], \left[\frac{x + y}{2}\right] - \left[\frac{x - y}{2}\right]\right) \end{aligned}$$

for all $x, y \in \mathbb{R}$. The mapping g satisfies (DOPP) but is not an isometry. Readers are encouraged to perform the calculations in detail as part of their exercises.

In the general case $\mathbb{R}^n$ for $n > 2$, the rotation does not work as in $\mathbb{R}^2$. This happens because the balls in the metrics d_m and d_σ have the same shape only for $n \in \{1, 2\}$. In $\mathbb{R}^2$, the balls are squares in both cases, but in $\mathbb{R}^3$ the balls are cubes for d_m and octahedrons for d_σ.

Example 4.10 Let n be an integer greater than 2. We define a mapping $f : (\mathbb{R}^n, d_m) \to (\mathbb{R}^n, d_m)$ by $f(x_1, x_2, \ldots, x_n) = ([x_1], [x_2], \ldots, [x_n])$.

Let $C_{i_1,i_2,\ldots,i_n}$ denote an n-cube defined by

$$C_{i_1,i_2,\ldots,i_n} = \big\{(x_1, x_2, \ldots, x_n) \in \mathbb{R}^n : i_k \leq x_k < i_k + 1 \\ \text{for all } k \in \{1, 2, \ldots, n\}\big\} \tag{4.1}$$

for all $i_1, i_2, \ldots, i_n \in \mathbb{Z}$. Then, we have

$$\mathbb{R}^n = \bigcup_{i_1 \in \mathbb{Z}} \bigcup_{i_2 \in \mathbb{Z}} \cdots \bigcup_{i_n \in \mathbb{Z}} C_{i_1,i_2,\ldots,i_n}.$$

Due to the definition of the mapping f and by (4.1), it follows that

$$\begin{aligned} f(x) &= f(x_1, x_2, \ldots, x_n) \\ &= \big([x_1], [x_2], \ldots, [x_n]\big) \\ &= (i_1, i_2, \ldots, i_n) \end{aligned} \tag{4.2}$$

for all $x = (x_1, x_2, \ldots, x_n) \in C_{i_1,i_2,\ldots,i_n}$.

If $x, y \in \mathbb{R}^n$ with $d_m(x, y) = 1$, then there are $i_1, i_2, \ldots, i_n, j_1, j_2, \ldots, j_n \in \mathbb{Z}$ with the following properties:

(i) $x \in C_{i_1,i_2,\ldots,i_n}$;
(ii) $y \in C_{j_1,j_2,\ldots,j_n}$;
(iii) There exists a nonempty subset I of $\{1, 2, \ldots, n\}$ such that

$$|i_k - j_k| = \begin{cases} 1 \text{ (for } k \in I), \\ 0 \text{ (otherwise)}. \end{cases}$$

Thus, it follows from (4.2) and (iii) that

$$d_m\big(f(x), f(y)\big) = \max_{k \in \{1,2,\ldots,n\}} |i_k - j_k| = 1$$

for all $x, y \in \mathbb{R}^n$ with $d_m(x, y) = 1$, which implies that f satisfies (DOPP) with respect to the metric d_m.

On the other hand, in view of the definition of the mapping f, it is easy to show that f is not an isometry. Therefore, the mapping f satisfies (DOPP) but is not an isometry.

We have not yet constructed an example for the metric d_σ like Example 4.10, nor have we proved that each mapping satisfying (DOPP) with respect to the metric d_σ is an isometry. So for $n > 2$ the following problem still remains unsolved:

Does every mapping $f : (\mathbb{R}^n, d_\sigma) \to (\mathbb{R}^n, d_\sigma)$ that satisfies (DOPP) *have to be an isometry?*

In case we could construct an appropriate non-isometric mapping that satisfies (DOPP), we could express $\mathbb{R}^n$ as the union of unit balls (which intersect only at boundary points). This is related to the partition hypothesis that applies to the relationship of the two conditions given in the problem below (see [13]).

Problem 4.11 What is the relationship between the following conditions?

(i) There exists a mapping $f : (\mathbb{R}^n, d) \to (\mathbb{R}^n, d)$ that satisfies (DOPP) but is not an isometry.
(ii) There exists a partition of $\mathbb{R}^n$ into unit balls with respect to the metric d having only boundary points in common.

Let us consider a surjective mapping $f : \mathbb{R}^2 \to \mathbb{R}^2$ whose domain and range have different metrics.

Theorem 4.12 *There is no mapping $f : (\mathbb{R}^2, d_m) \to (\mathbb{R}^2, d_e)$ that satisfies* (DOPP).

Proof Set $A = \{(0, 0), (1, 0), (0, 1), (1, 1)\}$. If $x \neq y$ for $x, y \in A$, then $d_m(x, y) = 1$. It then follows from (DOPP) that $d_e(f(x), f(y)) = 1$ for all $x, y \in A$ with $x \neq y$, which is a contradiction, since every subset of $\mathbb{R}^2$ (with metric d_e) having this property consists of at most three points. We have thus proved that no mapping $f : (\mathbb{R}^2, d_m) \to (\mathbb{R}^2, d_e)$ satisfies (DOPP). □

Now consider a mapping f, whose domain is endowed with the Euclidean metric d_e and whose range is endowed with the maximum metric d_m.

Theorem 4.13 *If there exists a non-isometric mapping $f : (\mathbb{R}^2, d_e) \to (\mathbb{R}^2, d_m)$ that satisfies* (DOPP)*, then there exists a non-isometric mapping $F : (\mathbb{R}^2, d_e) \to (\mathbb{R}^3, d_e)$ that also satisfies* (DOPP).

Proof Assume that $f : (\mathbb{R}^2, d_e) \to (\mathbb{R}^2, d_m)$ is a non-isometric mapping that satisfies (DOPP). Let us define

$$g : \mathbb{R}^2 \to \mathbb{Z}^2 \text{ by } g(x, y) = ([x], [y]);$$

$$h : \mathbb{Z}^2 \to A \text{ by } h(k, l) = (k \text{ (mod 2)}, \ l \text{ (mod 2)}), \text{ where}$$

$$A = \{(0, 0), (1, 0), (0, 1), (1, 1)\};$$

$$\Phi : A \to B \text{ as the bijection, where } B \text{ is the set of four vertices of a unit simplex in } \mathbb{R}^3.$$

It follows that the mappings

$$g : (\mathbb{R}^2, d_m) \to (\mathbb{Z}^2, d_m), \quad h : (\mathbb{Z}^2, d_m) \to (A, d_m), \quad \Phi : (A, d_m) \to (B, d_e)$$

satisfy (DOPP). Then the mapping $F : (\mathbb{R}^2, d_e) \to (\mathbb{R}^3, d_e)$ defined by

$$F = \Phi \circ h \circ g \circ f : \mathbb{R}^2 \to B \subset \mathbb{R}^3$$

satisfies (DOPP), but of course it is not an isometry. □

The aforementioned theorem illustrates an interesting consequence of considering different metrics in domain and range in the context of the classical problem mentioned in front of Example 4.4. A similar consequence can be obtained if we consider mappings between $(\mathbb{R}^2, d_\sigma)$ and $(\mathbb{R}^2, d_e)$.

Considering metrics other than the Euclidean metric, we obtain the following theorem based on previous considerations. When the Euclidean metric is applied in the following theorem, the mapping that satisfies (DOPP) is an isometry due to the Beckman-Quarles theorem.

Theorem 4.14 (Ciesielski and Rassias) *Let n be a fixed integer greater than 1. A continuous mapping $f : (\mathbb{R}^n, d_m) \to (\mathbb{R}^n, d_m)$ satisfying* (DOPP) *need not be an isometry.*

Proof Let us define a continuous mapping $f : (\mathbb{R}^2, d_m) \to (\mathbb{R}^2, d_m)$ by

$$f(x, y) = \big([x] + \{x\}^2, \ [y] + \{y\}^2\big),$$

where $\{x\} = x - [x]$ and $\{y\} = y - [y]$. We note that

$$
\begin{aligned}
&d_m\big(f(x, y), f(x', y')\big) \\
&\quad = d_m\big(([x] + \{x\}^2, [y] + \{y\}^2), ([x'] + \{x'\}^2, [y'] + \{y'\}^2)\big) \\
&\quad = \max\big\{\big|[x] + \{x\}^2 - [x'] - \{x'\}^2\big|, \big|[y] + \{y\}^2 - [y'] - \{y'\}^2\big|\big\}
\end{aligned}
$$

for all $(x, y), (x', y') \in \mathbb{R}^2$. When $d_m((x, y), (x', y')) = 1$, we may assume that $|x - x'| = 1$ and $|y - y'| \leq 1$ without loss of generality. In this case, we have either $[x] = [x'] - 1$ and $\{x\} = \{x'\}$ or $[x] = [x'] + 1$ and $\{x\} = \{x'\}$. Hence, f satisfies (DOPP) with respect to the maximum metric but it is not an isometry. This example proves our theorem for $n = 2$. By the same construction, we get examples for any integer $n > 2$. □

The following theorem can be proven using the rotation shown in Example 4.9. We encourage readers to prove this theorem themselves as an exercise.

Theorem 4.15 *A continuous mapping $f : (\mathbb{R}^2, d_\sigma) \to (\mathbb{R}^2, d_\sigma)$ that satisfies* (DOPP) *need not be an isometry.*

The following problem is still open: Assume that a continuous mapping $f : (\mathbb{R}^n, d_\sigma) \to (\mathbb{R}^n, d_\sigma)$ satisfies (DOPP) for an integer $n > 2$. Does f have to be an isometry?

4.4 Aleksandrov–Rassias Problems with (SDOPP)

In this section, we will investigate the properties of mappings that satisfy (SDOPP). Indeed, the content of this section is based on the paper [59] by Th. M. Rassias and P. Šemrl and follows its notations whenever possible.

We say that f preserves the distance n *in both directions* if for all $x, y \in X$, $\|x - y\| = n$ if and only if $\|f(x) - f(y)\| = n$. This is a concept similar to (SDOPP) defined in Definition 4.1 (ii).

Theorem 4.16 (Rassias and Šemrl) *Let X and Y be real normed spaces such that* $\dim X > 1$ *or* $\dim Y > 1$. *If $f : X \to Y$ is a surjective mapping that satisfies* (SDOPP), *then f is an one-to-one mapping such that*

$$
\big|\|f(x) - f(y)\| - \|x - y\|\big| < 1 \tag{4.3}
$$

for all $x, y \in X$. Moreover, f preserves distance n in both directions for any $n \in \mathbb{N}$.

Proof

(*a*) We prove that both spaces have dimension greater than 1. If $\dim Y > 1$, then there exist elements $x, y, z \in Y$ such that

$$\|x - y\| = \|y - z\| = \|z - x\| = 1.$$

Since the mapping f is assumed to be surjective and preserve the distance 1 in both directions, there are $x_1, y_1, z_1 \in X$ that satisfy

$$\|x_1 - y_1\| = \|y_1 - z_1\| = \|z_1 - x_1\| = 1.$$

This implies that $\dim X > 1$. Likewise, we may prove that if $\dim X > 1$, then $\dim Y > 1$.

(*b*) We claim that f is one-to-one. On the contrary, suppose there are $x, y \in X$ with $x \neq y$ such that $f(x) = f(y)$. Since $\dim X > 1$, we can choose a $z \in X$ such that $\|x - z\| = 1$ and $\|y - z\| \neq 1$. Because of our assumption that $f(y) = f(x)$, we have $\|f(y) - f(z)\| = \|f(x) - f(z)\| = 1$, where the last equality follows from (SDOPP). Furthermore, it follows from (SDOPP) that $\|y - z\| = 1$, which leads to a contradiction. Hence, we conclude that f is one-to-one. Therefore, we conclude that f is a bijective mapping and both f and f^{-1} preserve the unit distance.

(*c*) From now on, we need the following notations:

$$\overline{B}_r(x) = \{z : \|z - x\| \leq r\},$$
$$B_r(x) = \{z : \|z - x\| < r\},$$
$$C_{(n,n+1]}(x) = \{z : n < \|z - x\| \leq n + 1\}.$$

Let x be any element of X and $n > 1$ any integer. Assume that $z \in \overline{B}_n(x)$. Because of $\dim X > 1$, there exists a sequence $x = x_0, x_1, \ldots, x_n = z$ with

$$\|x_{i+1} - x_i\| = 1$$

for any $i \in \{0, 1, \ldots, n - 1\}$. Consequently, we have

$$\begin{aligned}\|f(z) - f(x)\| &= \|f(x_n) - f(x_0)\| \\ &\leq \sum_{i=0}^{n-1} \|f(x_{i+1}) - f(x_i)\| \\ &= n.\end{aligned}$$

Hence, we obtain

$$f\big(\overline{B}_n(x)\big) \subset \overline{B}_n\big(f(x)\big).$$

A similar result can be obtained for f^{-1}. Therefore, we conclude that

$$f\big(\overline{B}_n(x)\big) = \overline{B}_n\big(f(x)\big).$$

for all $x \in X$ and integers $n > 1$. However, f is bijective and thus

$$f\big(C_{(n,n+1]}(x)\big) = C_{(n,n+1]}\big(f(x)\big) \tag{4.4}$$

for all $x \in X$ and integers $n > 1$.

(*d*) We fix an element $x \in X$ and choose $z \in C_{(1,2]}(x)$. Then, it follows from (4.4) that $f(z) \in \overline{B}_2(f(x))$. We set $u = z + \frac{1}{\|z-x\|}(z-x)$. Then $\|u-z\| = 1$ and $u \in C_{(2,3]}(x)$. According to (4.4), we have $f(u) \in C_{(2,3]}(f(x))$. Thus, we obtain

$$\|f(u) - f(x)\| > 2. \tag{4.5}$$

If $\|f(z) - f(x)\| \leq 1$, then

$$\|f(u) - f(x)\| \leq \|f(u) - f(z)\| + \|f(z) - f(x)\| \leq \|f(u) - f(z)\| + 1 = 2,$$

which contradicts (4.5). Hence, we have proved that

$$f\big(C_{(1,2]}(x)\big) \subset C_{(1,2]}\big(f(x)\big).$$

The same result holds for the mapping f^{-1}. Consequently, the relations

$$f\big(C_{(1,2]}(x)\big) = C_{(1,2]}\big(f(x)\big) \quad \text{and} \quad f\big(B_1(x)\big) = B_1\big(f(x)\big)$$

hold for all $x \in X$. This together with (4.4) implies the validity of inequality (4.3).

(*e*) To complete the proof, we will show by induction on n that f preserves the distance n in both directions for all $n \in \mathbb{N}$. Assume that f preserves the distance n in both directions. Let x and z be elements of X such that $\|z - x\| = n + 1$. According to (4.4), we have $\|f(z) - f(x)\| \leq n + 1$. Let us define v by

$$v = f(x) + \frac{1}{\|f(z) - f(x)\|}\big(f(z) - f(x)\big).$$

Since f is surjective, there is a $u \in X$ such that $v = f(u)$. From $\|v - f(x)\| = 1$ we obtain $\|u - x\| = 1$ by (SDOPP).

If $\|v - f(z)\| < n$, it then follows from (4.4) that $\|u - z\| < n$. This fact together with $\|u - x\| = 1$ implies that

$$\|z - x\| \leq \|z - u\| + \|u - x\| < n + 1,$$

which contradicts our assumption that $\|z - x\| = n + 1$. Thus, it necessarily follows that $\|v - f(z)\| \geq n$. This implies

$$\begin{aligned} n &\leq \|v - f(z)\| \\ &= \left\| \left(1 - \frac{1}{\|f(z) - f(x)\|}\right) f(x) - \left(1 - \frac{1}{\|f(z) - f(x)\|}\right) f(z) \right\| \\ &= \big| \|f(z) - f(x)\| - 1 \big|. \end{aligned}$$

Therefore, we can conclude that $\|f(z) - f(x)\| = n + 1$. A similar proof also confirms that f^{-1} preserves the distance $n + 1$. □

Remark 4.17 The assumption that one of the spaces has dimension greater than 1 cannot be omitted in Theorem 4.16.

Proof If we define the mapping $f : \mathbb{R} \to \mathbb{R}$ by

$$f(x) = \begin{cases} x + 2 & (\text{for } x \in \mathbb{Z}), \\ x & (\text{for } x \notin \mathbb{Z}), \end{cases}$$

then f is a bijective mapping that preserves distance n in both directions for any $n \in \mathbb{N}$. However, since $f(0) = 2$ and $f(\frac{1}{3}) = \frac{1}{3}$, we have

$$\left\| f\left(\frac{1}{3}\right) - f(0) \right| - \left| \frac{1}{3} - 0 \right\| = \frac{4}{3} > 1.$$

Therefore, f does not fulfill inequality (4.3). □

Remark 4.18 In Theorem 4.16, (SDOPP) cannot be replaced by (DOPP).

Proof Let $g : [0, 1) \to [0, 1) \times \mathbb{R}$ be a bijective mapping. Furthermore, we define $f : \mathbb{R} \to \mathbb{R}^2$ by

$$f(t) = g(\{t\}) + ([t], 0),$$

where we set $\{t\} = t - [t]$, and we denote by $[t]$ the integer part of t.

We claim that f is a one-to-one mapping. Assume that t and t' are real numbers with $[t] \neq [t']$ and $\{t\} \neq \{t'\}$. Then we have

$$\begin{aligned} f(t) - f(t') &= g(\{t\}) - g(\{t'\}) + ([t] - [t'], 0) \\ &\in \left\{(x, y) \in \mathbb{R}^2 : [t] - [t'] - 1 < x < [t] - [t'] + 1,\ y \in \mathbb{R}\right\} \\ &\not\ni (0, 0). \end{aligned}$$

Suppose t and t' are real numbers with $[t] \neq [t']$ and $\{t\} = \{t'\}$. Then, we see that

$$f(t) - f(t') = ([t] - [t'], 0) \neq (0, 0).$$

Finally, if t and t' are real numbers with $[t] = [t']$ and $\{t\} \neq \{t'\}$, then

$$f(t) - f(t') = g(\{t\}) - g(\{t'\}) \neq (0, 0).$$

Hence, f is a one-to-one mapping.

Now we claim that f is a surjective mapping. Assume that (x, y) be an arbitrary point in $\mathbb{R}^2$. Since $g : [0, 1) \to [0, 1) \times \mathbb{R}$ is a bijective mapping, we can choose a real number t such that $x - 1 < [t] \leq x$ and

$$g(\{t\}) = (x - [t], y) = (x, y) - ([t], 0),$$

i.e., $f(t) = g(\{t\}) + ([t], 0) = (x, y)$, which implies that f is a surjective mapping.

Moreover, f preserves the unit distance:

$$\begin{aligned} \|f(t + 1) - f(t)\| &= \left\|g(\{t + 1\}) + ([t + 1], 0) - g(\{t\}) - ([t], 0)\right\| \\ &= \|(1, 0)\| \\ &= 1 \end{aligned}$$

for $t \in \mathbb{R}$, i.e., f satisfies (DOPP). However, f does not satisfy inequality (4.3). □

Remark 4.19 The inequality (4.3) is sharp.

Proof

(*a*) Given $\varepsilon \in (0, \frac{1}{2})$, we define a mapping $g_\varepsilon : [0, 1] \to [0, 1]$ by

$$g_\varepsilon(t) = \begin{cases} \frac{\varepsilon}{1-\varepsilon} t & (\text{for } t \in [0, 1 - \varepsilon]), \\ \frac{1-\varepsilon}{\varepsilon} t + \left(2 - \frac{1}{\varepsilon}\right) & (\text{for } t \in [1 - \varepsilon, 1]). \end{cases}$$

We note that g_ε is a strictly increasing mapping. We also define $h_\varepsilon : \mathbb{R} \to \mathbb{R}$ by

$$h_\varepsilon(s) = [s] + g_\varepsilon(\{s\}),$$

where we set $\{s\} = s - [s]$.

We assert that h_ε is a strictly increasing mapping. We note that for any real numbers s and t, $s > t$ if and only if either $[s] > [t]$ or $[s] = [t]$ and $\{s\} > \{t\}$. Assume that $[s] > [t]$. Then we obtain

$$h_\varepsilon(s) - h_\varepsilon(t) = \big([s] - [t]\big) + \big(g_\varepsilon(\{s\}) - g_\varepsilon(\{t\})\big) > 0,$$

since $0 < g_\varepsilon(\{s\}) - g_\varepsilon(\{t\}) < 1$. Now we assume that $[s] = [t]$ and $\{s\} > \{t\}$. Then we have

$$h_\varepsilon(s) - h_\varepsilon(t) = g_\varepsilon(\{s\}) - g_\varepsilon(\{t\}) > 0,$$

since g_ε is a strictly increasing mapping. Thus, h_ε is a strictly increasing mapping.

(b) Since $-1 < g_\varepsilon(\{s\}) - g_\varepsilon(\{t\}) < 1$ for all $s, t \in \mathbb{R}$, it holds that for $n \in \mathbb{N}_0$,

$$|h_\varepsilon(s) - h_\varepsilon(t)| = \big|\big([s] - [t]\big) + \big(g_\varepsilon(\{s\}) - g_\varepsilon(\{t\})\big)\big| = n$$

if and only if $g_\varepsilon(\{s\}) = g_\varepsilon(\{t\})$, which is equivalent to the fact that $\{s\} = \{t\}$. In this case, we remark that $s - t = [s] + \{s\} - [t] - \{t\} = [s] - [t] = \pm n$. Hence, we may conclude that

$$|s - t| = n \quad \text{if and only if} \quad |h_\varepsilon(s) - h_\varepsilon(t)| = n$$

for each $n \in \mathbb{N}$.

(c) Without loss of generality, assume that s and t are real numbers with $s > t$. Similarly, we can verify that if $|s - t| \le n$ for some $n \in \mathbb{N}$, then $|h_\varepsilon(s) - h_\varepsilon(t)| \le n$. The aforementioned proof is not difficult, so readers are encouraged to solve it themselves.

We now concentrate on proving the converse. Assume that $|h_\varepsilon(s) - h_\varepsilon(t)| \le n$ for some $n \in \mathbb{N}$. If s and t are arbitrary real numbers with $s > t$ and $[s] = [t]$, then $\{s\} > \{t\}$. Hence, we have

$$|s - t| = [s] - [t] + \{s\} - \{t\} = \{s\} - \{t\} < 1 \le n.$$

Assume that $[s] - [t] \in \{1, 2, \ldots, n - 1\}$ for some $s, t \in \mathbb{R}$ with $s > t$. Since $-1 < \{s\} - \{t\} < 1$, we obtain

$$|s - t| = [s] - [t] + \{s\} - \{t\} < n - 1 + 1 = n.$$

Finally, suppose $[s] - [t] = n$. From our assumption that

$$|h_\varepsilon(s) - h_\varepsilon(t)| = [s] - [t] + g_\varepsilon(\{s\}) - g_\varepsilon(\{t\}) \leq n,$$

it follows that $g_\varepsilon(\{s\}) \leq g_\varepsilon(\{t\})$, which implies that $\{s\} \leq \{t\}$. Thus, we get

$$|s - t| = [s] - [t] + \{s\} - \{t\} \leq n.$$

Consequently, we have showed that

$$|s - t| \leq n \quad \text{if and only if} \quad |h_\varepsilon(s) - h_\varepsilon(t)| \leq n$$

for any $n \in \mathbb{N}$.

(d) Now we assert that

$$|h_\varepsilon(s) - h_\varepsilon(t)| \leq \frac{1-\varepsilon}{\varepsilon}|s - t| \tag{4.6}$$

for all $s, t \in \mathbb{R}$. Without loss of generality, we assume that $s > t$, which is equivalent to the fact that either $[s] > [t]$ or $[s] = [t]$ and $\{s\} > \{t\}$. If $[s] > [t]$, then

$$\begin{aligned}
&|h_\varepsilon(s) - h_\varepsilon(t)| - \frac{1-\varepsilon}{\varepsilon}|s - t| \\
&= h_\varepsilon(s) - h_\varepsilon(t) - \frac{1-\varepsilon}{\varepsilon}(s - t) \\
&= [s] - [t] + \big(g_\varepsilon(\{s\}) - g_\varepsilon(\{t\})\big) - \frac{1-\varepsilon}{\varepsilon}(s - t) \\
&\leq [s] - [t] + \frac{1-\varepsilon}{\varepsilon}\{s\} + 2 - \frac{1}{\varepsilon} - \frac{\varepsilon}{1-\varepsilon}\{t\} - \frac{1-\varepsilon}{\varepsilon}\big([s] + \{s\} - [t] - \{t\}\big) \\
&= \left(2 - \frac{1}{\varepsilon}\right)\big([s] - [t]\big) + 2 - \frac{1}{\varepsilon} + \left(\frac{1}{\varepsilon} - 1 - \frac{\varepsilon}{1-\varepsilon}\right)\{t\} \\
&= \left(2 - \frac{1}{\varepsilon}\right)\big([s] - [t] + 1 - \{t\}\big) + \left(1 - \frac{\varepsilon}{1-\varepsilon}\right)\{t\} \\
&< 2 - \frac{1}{\varepsilon} + 1 - \frac{\varepsilon}{1-\varepsilon} \\
&< 0
\end{aligned}$$

for any $\varepsilon \in (0, \frac{1}{2})$.

We now consider the other case when $[s] = [t]$ and $\{s\} > \{t\}$. If $0 \leq \{t\} < \{s\} \leq 1 - \varepsilon$, then

$$\begin{aligned}
&|h_\varepsilon(s) - h_\varepsilon(t)| - \frac{1-\varepsilon}{\varepsilon}|s-t| \\
&= g_\varepsilon(\{s\}) - g_\varepsilon(\{t\}) - \frac{1-\varepsilon}{\varepsilon}(s-t) \\
&\le \frac{\varepsilon}{1-\varepsilon}\{s\} - \frac{\varepsilon}{1-\varepsilon}\{t\} - \frac{1-\varepsilon}{\varepsilon}\big([s] + \{s\} - [t] - \{t\}\big) \\
&= \left(\frac{\varepsilon}{1-\varepsilon} - \frac{1-\varepsilon}{\varepsilon}\right)\big(\{s\} - \{t\}\big) \\
&< 0
\end{aligned}$$

for any $\varepsilon \in (0, \frac{1}{2})$.

If $[s] = [t]$ and $0 \le \{t\} \le 1 - \varepsilon < \{s\} \le 1$, then

$$\begin{aligned}
&|h_\varepsilon(s) - h_\varepsilon(t)| - \frac{1-\varepsilon}{\varepsilon}|s-t| \\
&= g_\varepsilon(\{s\}) - g_\varepsilon(\{t\}) - \frac{1-\varepsilon}{\varepsilon}(s-t) \\
&\le \frac{1-\varepsilon}{\varepsilon}\{s\} + 2 - \frac{1}{\varepsilon} - \frac{\varepsilon}{1-\varepsilon}\{t\} - \frac{1-\varepsilon}{\varepsilon}\big([s] + \{s\} - [t] - \{t\}\big) \\
&= 2 - \frac{1}{\varepsilon} + \left(\frac{1-\varepsilon}{\varepsilon} - \frac{\varepsilon}{1-\varepsilon}\right)\{t\} \\
&\le 2 - \frac{1}{\varepsilon} + \left(\frac{1-\varepsilon}{\varepsilon} - \frac{\varepsilon}{1-\varepsilon}\right)(1-\varepsilon) \\
&= 0
\end{aligned}$$

for all $\varepsilon \in (0, \frac{1}{2})$.

If $[s] = [t]$ and $1 - \varepsilon < \{t\} < \{s\} \le 1$, then

$$\begin{aligned}
&|h_\varepsilon(s) - h_\varepsilon(t)| - \frac{1-\varepsilon}{\varepsilon}|s-t| \\
&= g_\varepsilon(\{s\}) - g_\varepsilon(\{t\}) - \frac{1-\varepsilon}{\varepsilon}(s-t) \\
&= \frac{1-\varepsilon}{\varepsilon}\{s\} - \frac{1-\varepsilon}{\varepsilon}\{t\} - \frac{1-\varepsilon}{\varepsilon}\big(\{s\} - \{t\}\big) \\
&= 0
\end{aligned}$$

for any $\varepsilon \in (0, \frac{1}{2})$. Therefore, we have proved that inequality (4.6) holds for all $s, t \in \mathbb{R}$ and $\varepsilon \in (0, \frac{1}{2})$.

(e) We use the notation $C[0, 1]$ to denote the set of all real-valued continuous mappings defined on $[0, 1]$. We introduce the norm $\|x\|_m = \max_{t\in[0,1]} |x(t)|$ on $C[0, 1]$, where $x \in C[0, 1]$. In addition, we define the mapping $\phi_\varepsilon : C[0, 1] \to C[0, 1]$ by

$$\big(\phi_\varepsilon(x)\big)(t) = h_\varepsilon\big(x(t)\big)$$

for all $x \in C[0, 1]$ and $t \in [0, 1]$. Obviously, ϕ_ε is a bijective mapping, and we see that $(\phi_\varepsilon^{-1}(x))(t) = h_\varepsilon^{-1}(x(t))$.

Moreover, for any pair $x, y \in C[0, 1]$ the following conditions are equivalent:

(i) $\|x - y\|_m = n$.

(ii) There exists $t_0 \in [0, 1]$ such that $|x(t_0) - y(t_0)| = n$ and $|x(t) - y(t)| \le n$ for all $t \in [0, 1]$.

(iii) $|h_\varepsilon(x(t_0)) - h_\varepsilon(y(t_0))| = n$ and $|h_\varepsilon(x(t)) - h_\varepsilon(y(t))| \le n$ for $t \in [0, 1]$.

(iv) $\|\phi_\varepsilon(x) - \phi_\varepsilon(y)\|_m = n$.

In view of the definition of the norm $\|\cdot\|_m$, conditions (i) and (ii) are equivalent. Due to (b) and (c), the conditions in (ii) and (iii) are equivalent. By the definition of ϕ_ε, (iii) and (iv) are obviously equivalent. Therefore, the mapping ϕ_ε preserves distance n in both directions for any $n \in \mathbb{N}$.

(f) We define $x, y \in C[0, 1]$ by $x(t) = 1 - \varepsilon$ and $y(t) = 1$ for all $t \in [0, 1]$. Then $\|x - y\|_m = \varepsilon$. In addition, we see that

$$\big(\phi_\varepsilon(x)\big)(t) = h_\varepsilon\big(x(t)\big) = h_\varepsilon(1 - \varepsilon) = [1 - \varepsilon] + g_\varepsilon(\{1 - \varepsilon\}) = \varepsilon$$

and

$$\big(\phi_\varepsilon(y)\big)(t) = h_\varepsilon\big(y(t)\big) = h_\varepsilon(1) = [1] + g_\varepsilon(\{1\}) = 1$$

for all $t \in [0, 1]$. It thus follows that $\|\phi_\varepsilon(x) - \phi_\varepsilon(y)\|_m = 1 - \varepsilon$. Consequently, we obtain

$$\big|\|\phi_\varepsilon(x) - \phi_\varepsilon(y)\|_m - \|x - y\|_m\big| = 1 - 2\varepsilon,$$

and since we can choose ε as small as we like, the inequality (4.3) is sharp. □

The mapping $\phi_\varepsilon : C[0, 1] \to C[0, 1]$ in Remark 4.19 is not only continuous, but also satisfies a stronger condition (Lipschitz condition):

$$\begin{aligned}
\|\phi_\varepsilon(x) - \phi_\varepsilon(y)\|_m &= \max_{t\in[0,1]} \big|\big(\phi_\varepsilon(x) - \phi_\varepsilon(y)\big)(t)\big| \\
&= \max_{t\in[0,1]} \big|h_\varepsilon\big(x(t)\big) - h_\varepsilon\big(y(t)\big)\big| \\
&\le \frac{1-\varepsilon}{\varepsilon} \max_{t\in[0,1]} |x(t) - y(t)| \\
&= K\|x - y\|_m
\end{aligned}$$

for any $x, y \in C[0, 1]$, where the inequality sign is due to (4.6) and we set $K = \frac{1-\varepsilon}{\varepsilon} > 1$. It follows that surjective Lipschitz mappings ($K > 1$) that satisfy (SDOPP) need not be isometries. However, for the case that $K = 1$, we can prove an extension of the Beckman-Quarles theorem.

Theorem 4.20 (Rassias and Šemrl) *Let X and Y be real normed spaces such that $\dim X > 1$ or $\dim Y > 1$. Assume that $f : X \to Y$ is a Lipschitz mapping with a Lipschitz constant* 1, *i.e.,*

$$\|f(x) - f(y)\| \le \|x - y\|$$

for all $x, y \in X$. If f is a surjective mapping satisfying (SDOPP), *then f is an affine isometry.*

Proof According to Theorem 4.16, f preserves the distance n in both directions for any $n \in \mathbb{N}$. We can choose two different elements $x, y \in X$ and an integer $n_0 > 0$ such that $\|x - y\| < n_0$. Assume that

$$\|f(x) - f(y)\| < \|x - y\|. \tag{4.7}$$

If we set

$$z = x - \frac{n_0}{\|x - y\|}(x - y),$$

then $\|z - x\| = n_0$ and

$$\|z - y\| = \left\| \left(1 - \frac{n_0}{\|x - y\|}\right)(x - y) \right\| = n_0 - \|x - y\|.$$

Since f preserves distance n_0 in both directions, it follows from the Lipschitz condition that

$$\|f(z) - f(x)\| = n_0 \quad \text{and} \quad \|f(z) - f(y)\| \le n_0 - \|x - y\|. \tag{4.8}$$

On the other hand, by (4.7) and (4.8), we obtain

$$\begin{aligned} \|f(z) - f(x)\| &\le \|f(z) - f(y)\| + \|f(y) - f(x)\| \\ &< n_0 - \|x - y\| + \|x - y\| \\ &= n_0, \end{aligned}$$

which contradicts (4.8). Hence, the strict inequality (4.7) cannot hold. Therefore, we conclude that

$$\|f(x) - f(y)\| = \|x - y\|$$

for all $x, y \in X$, which implies that f is an isometry. Finally, we use the theorem of Mazur and Ulam to prove that f is an affine isometry. □

Let X and Y be real normed spaces such that $\dim X > 1$ and Y is strictly convex. It follows from the theorem of Benz and Berens that if $f : X \to Y$ preserves two distances ρ and $N\rho$ (where $N > 1$ is an integer and $\rho > 0$ is a real number), then f is an affine isometry.

Corollary 4.21 *Let X and Y be real normed spaces such that* $\dim X > 1$ *or* $\dim Y > 1$. *Also assume that one of them is strictly convex. If* $f : X \to Y$ *is a surjective mapping that satisfies* (SDOPP), *then* f *is an affine isometry.*

Proof According to Theorem 4.16, f is a bijective mapping that preserves the distance m in both directions for every $m \in \mathbb{N}$. Moreover, the proof of Theorem 4.16 shows that $\dim X > 1$ and $\dim Y > 1$. So, without loss of generality, we can assume that Y is strictly convex.

We first claim that f preserves the distance $\frac{1}{n}$ for every $n \in \mathbb{N}$: If we choose $x, y \in X$ with $\|x - y\| = \frac{1}{n}$, then there exists a vector $z \in X$ such that $\|x - z\| = \|y - z\| = 1$. Set

$$u = z + n(y - z) \quad \text{and} \quad v = z + n(x - z).$$

Obviously, we have

$$\|x - v\| = n - 1 \quad \text{and} \quad \|v - z\| = n.$$

Since f preserves the distance m in both directions for any $m \in \mathbb{N}$, we have

$$\|f(x) - f(z)\| = 1, \quad \|f(x) - f(v)\| = n - 1, \quad \|f(v) - f(z)\| = n.$$

Since Y is strictly convex, it necessarily follows that

$$f(x) = \frac{1}{n} f(v) + \frac{n-1}{n} f(z).$$

Similarly, we get

$$f(y) = \frac{1}{n} f(u) + \frac{n-1}{n} f(z).$$

Using $\|v - u\| = 1$, we obtain $\|f(x) - f(y)\| = \frac{1}{n}\|f(v) - f(u)\| = \frac{1}{n}$.

Now we assume that $\|x-y\| \leq \frac{m}{n}$, where m is an integer greater than 1. Since $\dim X > 1$, there is a finite sequence $\{z_0 = x, z_1, \ldots, z_m = y\}$ of vectors in X such that $\|z_i - z_{i+1}\| = \frac{1}{n}$ for $i \in \{0, 1, \ldots, m-1\}$. Hence, we have

$$\|f(x) - f(y)\| \leq \sum_{i=0}^{m-1} \|f(z_i) - f(z_{i+1})\| = \frac{m}{n}.$$

Using the ideas of (g) and (h) in the proof of Theorem 3.3, we conclude that

$$\|f(x) - f(y)\| \leq \|x - y\|$$

for all $x, y \in X$. We now apply Theorem 4.20 to complete the proof. □

4.5 Aleksandrov–Rassias Problems with (DOPP)

In the previous section, we investigated the properties of mappings that satisfy (SDOPP). In this section, we will examine the properties of mappings that satisfy (DOPP).

The following result was proved in [51] by B. Mielnik and Th. M. Rassias under the additional assumption that f is a homeomorphism.

Theorem 4.22 (Mielnik and Rassias) *Suppose X and Y are real normed spaces such that* $\dim X > 1$, *and one of them is strictly convex. If $f : X \to Y$ is a homeomorphism satisfying* (DOPP), *then f is an affine isometry.*

Proof Let x be an arbitrary vector in X. Since f satisfies (DOPP), f maps the unit sphere $S_1(x) = \{z \in X : \|z - x\| = 1\}$ into the sphere $S_1(f(x)) = \{z \in Y : \|z - f(x)\| = 1\}$. Let Z be the complement of $S_1(x)$ in X and $\tilde{Z}$ the complement of $f(S_1(x))$ in Y. While the homeomorphism f maps $S_1(x)$ onto $f(S_1(x))$, it must simultaneously map Z onto $\tilde{Z}$.

We claim that $f(S_1(x)) = S_1(f(x))$. On the contrary, suppose $f(S_1(x))$ is a proper subset of $S_1(f(x))$. Then its complement $\tilde{Z}$ is connected, which is impossible since $\tilde{Z} = f(Z)$ and Z is disconnected (see Remark 1.19). Thus, the unit sphere $S_1(x)$ is mapped onto the unit sphere $S_1(f(x))$. Therefore, the mapping f satisfies (SDOPP). Finally, we complete the proof using Corollary 4.21. □

Theorem 4.20 was further generalized by Th. M. Rassias and S. Xiang [60].

Theorem 4.23 (Rassias and Xiang) *Let X and Y be real normed spaces. Assume that Y is strictly convex and $f : X \to Y$ is a Lipschitz mapping with a Lipschitz constant* 1*:*

$$\|f(x) - f(y)\| \leq \|x - y\|$$

for all $x, y \in X$. If f satisfies (DOPP), *then f is an affine isometry.*

Proof

(*a*) Assume that x and y are arbitrary elements of X with $\|x - y\| = \frac{1}{2}$. If we set $z = x + 2(y - x)$, then we have $\|x - z\| = 1$ and $\|y - z\| = \frac{1}{2}$. Since f is a Lipschitz mapping with a Lipschitz constant 1 and satisfies (DOPP), we get

$$\begin{aligned}\frac{1}{2} &= \|x - y\| \\ &\geq \|f(x) - f(y)\| \\ &\geq \|f(x) - f(z)\| - \|f(y) - f(z)\| \\ &\geq 1 - \|y - z\| \\ &= \frac{1}{2}.\end{aligned}$$

Hence, we have $\|f(x) - f(y)\| = \frac{1}{2}$. Therefore, f preserves the distances 1 and $\frac{1}{2}$.

(*b*) Applying a similar method as in the proof of Theorem 3.3, it is easy to verify that f preserves the distance $n + \frac{1}{2}$ in both directions for any $n \in \mathbb{N}$. We invite the reader to try out this proof.

(*c*) We now choose arbitrary elements x, y of X and an $n_0 \in \mathbb{N}$ such that $\|x - y\| < n_0 + \frac{1}{2}$. Assume that

$$\|f(x) - f(y)\| < \|x - y\|. \tag{4.9}$$

If we set

$$z = x + \frac{2n_0 + 1}{2\|x - y\|}(y - x),$$

then we have

$$\|z - x\| = n_0 + \frac{1}{2} \quad \text{and} \quad \|z - y\| = n_0 + \frac{1}{2} - \|y - x\|.$$

Moreover, since f preserves the distances $n + \frac{1}{2}$, $n \in \mathbb{N}$, by (*b*) and f is a Lipschitz mapping with a Lipschitz constant 1, it holds that

$$\|f(z) - f(x)\| = n_0 + \frac{1}{2} \quad \text{and} \quad \|f(z) - f(y)\| \leq n_0 + \frac{1}{2} - \|y - x\|.$$

On the other hand, it follows from (4.9) that

$$\begin{aligned} n_0 + \frac{1}{2} &= \|f(z) - f(x)\| \\ &\leq \|f(z) - f(y)\| + \|f(y) - f(x)\| \\ &< n_0 + \frac{1}{2} - \|y - x\| + \|y - x\| \\ &= n_0 + \frac{1}{2}, \end{aligned}$$

which leads to a contradiction. Hence, the strict inequality (4.9) is not valid. Therefore, since f is a Lipschitz mapping with a Lipschitz constant 1, we conclude that $\|f(x) - f(y)\| = \|x - y\|$ for all $x, y \in X$, i.e., f is an isometry. Finally, by Baker's theorem, f is an affine isometry. □

We note that Y. Ma [48] proved the following theorem, which is more general than Theorem 4.23.

Theorem 4.24 (Ma) *Let X and Y be real normed spaces. Assume that Y is strictly convex and $f : X \to Y$ is a local Lipschitz mapping with a Lipschitz constant* 1*:*

$$\|f(x) - f(y)\| \leq \|x - y\|$$

for all $x, y \in X$ with $\|x - y\| \leq 1$. If f satisfies (DOPP)*, then f is an affine isometry.*

We assume that a mapping $f : \mathbb{E}^n \to \mathbb{E}^n$ satisfies (DOPP), where n is an integer greater than 1. According to part (a) of the proof of the Beckman-Quarles theorem presented at the end of Chap. 2 or Lemma 2.21, the mapping f preserves distance $\sqrt{2(1 + \frac{1}{n})}$.

In general, let X and Y be real Hilbert spaces with $\dim X \geq n$, where n is an integer greater than 1. We assert that if a mapping $f : X \to Y$ satisfies (DOPP), then $\|f(p) - f(q)\| \leq \sqrt{2(1 + \frac{1}{n})}$ for any $p, q \in X$ with $\|p - q\| = \sqrt{2(1 + \frac{1}{n})}$: Let $p, p_1, \ldots, p_n$ and $q, p_1, \ldots, p_n$ be the vertices of two unit regular n-simplices in X, respectively, and $\|p - q\| = \sqrt{2(1 + \frac{1}{n})}$.

We set $z = f(p) - f(q)$, $x_i = f(p) - f(p_i)$ and $y_i = f(p_i) - f(q)$ for each $i \in \{1, 2, \ldots, n\}$. Then we have

$$\|x_i\| = \|y_i\| = \|x_i - x_j\| = 1 \quad \text{and} \quad z = x_i + y_i$$

for all $i, j \in \{1, 2, \ldots, n\}$ with $i \neq j$. Since $\|z - x_i\|^2 = \|z - y_i\|^2 = 1$ for any $i \in \{1, 2, \ldots, n\}$, it follows from the Cauchy-Schwarz inequality that

$$\|z\|^2 = 2|\langle z, x_i \rangle| = \frac{2}{n}\left|\left\langle z, \sum_{i=1}^{n} x_i \right\rangle\right| \leq \frac{2}{n}\|z\|\left\|\sum_{i=1}^{n} x_i\right\| \leq \sqrt{2\left(1+\frac{1}{n}\right)}\,\|z\|.$$

Therefore, it holds that $\|f(p) - f(q)\| = \|z\| \leq \sqrt{2(1+\frac{1}{n})}$.

The following theorem is a consequence of the Benz–Berens theorem.

Theorem 4.25 *Let X and Y be real Hilbert spaces with* $\dim X \geq n$*, where n is an integer greater than* 1*. If a mapping* $f : X \to Y$ *satisfies* (DOPP) *and preserves the distance* $k\sqrt{2(1+\frac{1}{n})}$ *for some integer* $k > 1$*, then* f *is an affine isometry.*

In addition, many researchers have attempted to solve the Aleksandrov–Rassias problems several times, but have not yet succeeded in completely solving the problems.

Rassias and Xiang's Partial Solutions 5

Abstract

In this and the next chapters, we will introduce in more detail the ideas and methods used to solve the Aleksandrov–Rassias problems for each case. Section 5.1 focuses on examining the Aleksandrov–Rassias problem, where the relevant mapping preserves two distances, 1 and $\sqrt{3}$. This case is a special case among cases where the ratio of two distances is not an integer. In Sect. 5.2, we consider in detail the Aleksandrov–Rassias problem where the relevant mapping preserves the distances 1 and $\sqrt{2}$. Section 5.3 investigates the conditions under which the relevant mapping preserving three distances necessarily become an isometry. As shown in Problem 4.2 (ii), this problem is also closely related to the Aleksandrov–Rassias problems. In this chapter, we extract the main results from the papers by Rassias and Xiang (Univ Beograd Publ Elektrotehn Fak 11(4):1–8, 2000), Xiang (Aleksandrov problem and mappings which preserve distances, in *Functional Equations and Inequalities*, ed. by Th.M. Rassias, pp. 297–323, Kluwer, Alphen aan den Rijn, 2000), Xiang (J Math Anal Appl 254(1):262–274, 2001) and organize them so that readers can easily understand them.

5.1 The Case Where 1 and $\sqrt{3}$ Are Preserved

In this section, we consider the Aleksandrov–Rassias problem for the case that the mapping preserves two distances 1 and $\sqrt{3}$. This is a special case among cases where the ratio of two distances is not an integer.

The following theorem is excerpted from a paper [72] written by S. Xiang.

Theorem 5.1 (Xiang) *Let X and Y be real Hilbert spaces with* $\dim X > 1$. *If a mapping $f : X \to Y$ preserves the distances* 1 *and* $\sqrt{3}$, *then f is an affine isometry.*

S.-M. Jung, *Aleksandrov-Rassias Problems on Distance Preserving Mappings*,
Frontiers in Mathematics, https://doi.org/10.1007/978-3-031-77613-7_5

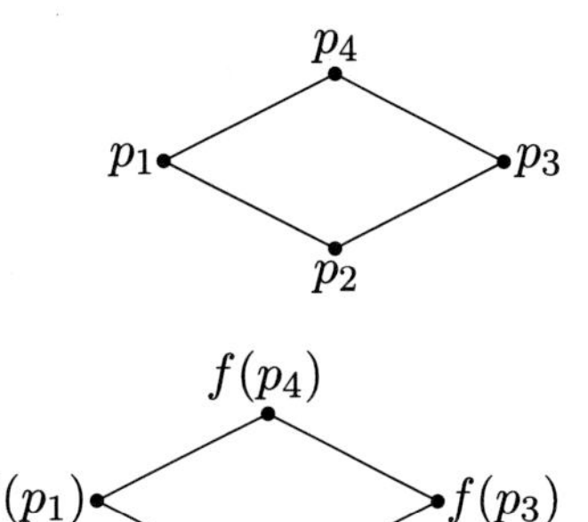

Fig. 5.1 A rhombus with sides of length 1 and diagonals of length $\sqrt{3}$ and 1 is drawn. The vertices of this rhombus are p_1, p_2, p_3, p_4

Fig. 5.2 The image of vertices of a rhombus under f is depicted, and their preimage was presented in Fig. 5.1

Proof

(*a*) Let p_1, p_2, p_3, p_4 be the points of X, which are the vertices of a rhombus with unit side length such that $\|p_1 - p_3\| = \sqrt{3}$ and $\|p_2 - p_4\| = 1$, as shown in Fig. 5.1.

(*b*) We now set $x = f(p_2) - f(p_1)$, $y = f(p_4) - f(p_1)$, and $z = f(p_3) - f(p_1)$. Since f preserves the distances 1 and $\sqrt{3}$, it holds that

$$\|x\| = \|y\| = \|x - y\| = \|z - x\| = \|z - y\| = 1 \quad \text{and} \quad \|z\| = \sqrt{3}.$$

Since Y is a real Hilbert space, we obtain

$$\|x - y\|^2 = \|x\|^2 - 2\langle x, y\rangle + \|y\|^2 = 1,$$
$$\|z - x\|^2 = \|z\|^2 - 2\langle z, x\rangle + \|x\|^2 = 1,$$
$$\|z - y\|^2 = \|z\|^2 - 2\langle z, y\rangle + \|y\|^2 = 1.$$

Thus, it follows that

$$\langle x, y\rangle = \frac{1}{2}, \quad \langle z, x\rangle = \langle z, y\rangle = \frac{3}{2},$$

and

$$\|z - x - y\|^2 = \|z\|^2 - 2\langle z, x\rangle - 2\langle z, y\rangle + \|x\|^2 + 2\langle x, y\rangle + \|y\|^2 = 0.$$

The previous equality implies that $z = x + y$, *i.e.*, $f(p_1), f(p_2), f(p_3), f(p_4)$ in Y are the vertices of a rhombus of unit side length such that $\|f(p_1) - f(p_3)\| = \sqrt{3}$ and $\|f(p_2) - f(p_4)\| = 1$. Moreover, $f(p_3)$ is in the span of $\{f(p_1), f(p_2), f(p_4)\}$. In particular, $f(p_3) - f(p_1) = (f(p_2) - f(p_1)) + (f(p_4) - f(p_1))$. (See Fig. 5.2.)

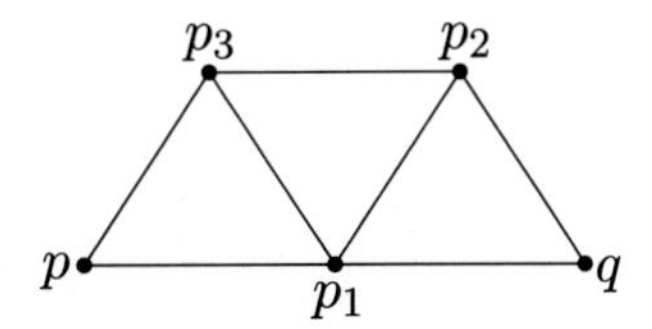

Fig. 5.3 Let p and q be two points whose distance from each other is 2, and let the midpoint of these points be p_1. Two more points p_2 and p_3 are chosen such that p, p_1, p_2, p_3 and p_1, q, p_2, p_3 are the vertices of two rhombi with side length 1

(*c*) Assume that p and q are arbitrary points of X with $\|p - q\| = 2$. We set $p_1 = \frac{1}{2}(p + q)$. Since $\dim X > 1$, there are points p_2 and p_3 of X such that p, p_1, p_2, p_3 and p_1, q, p_2, p_3 are the vertices of two rhombi of unit side length with $\|p_2 - p\| = \|p_3 - q\| = \sqrt{3}$ (see Fig. 5.3).
By (*b*), $f(p), f(p_1), f(p_2), f(p_3)$ and $f(p_1), f(q), f(p_2), f(p_3)$ are the vertices of two rhombi of unit side length with $\|f(p_2) - f(p)\| = \|f(p_3) - f(q)\| = \sqrt{3}$.
If we set $x = f(p_1) - f(p)$ and $y = f(p_3) - f(p)$, then $x - y = f(p_1) - f(p_3)$, $x + y = f(p_2) - f(p)$, and $f(q) - f(p_3) = (f(p_1) - f(p_3)) + (f(p_2) - f(p_3))$. Thus, we obtain

$$f(q) = \big(f(p_1) - f(p_3)\big) + f(p_2) = (x - y) + f(p_2).$$

Hence, we have

$$\begin{aligned} f(q) - f(p) &= (x - y) + \big(f(p_2) - f(p)\big) \\ &= (x - y) + (x + y) \\ &= 2x \\ &= 2\big(f(p_1) - f(p)\big) \end{aligned}$$

and $\|f(q) - f(p)\| = 2$. Therefore, f preserves the distance 2.

(*d*) Finally, it follows from Theorem 2.5 and the Benz–Berens theorem that f is an affine isometry. □

In Theorem 5.1, the condition that $\dim X > 1$ cannot be relaxed. For example, let $f : \mathbb{R} \to \mathbb{R}$ be a mapping of the form $f(x) = x + \phi(x)$, where ϕ is defined by

$$\phi(x) = \begin{cases} 0 \text{ (for } x \in A), \\ 1 \text{ (for } x \notin A), \end{cases}$$

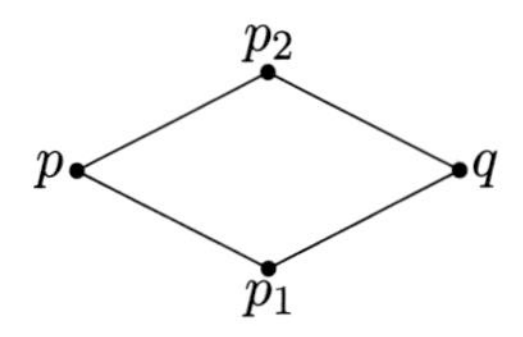

Fig. 5.4 Let p and q be two points whose distance from each other is $\sqrt{3}$. Two more points p_1 and p_2 are chosen such that p, p_1, q, p_2 become the vertices of a rhombus with side length 1

where $A = \{a + b\sqrt{3} \in \mathbb{R} : a, b \in \mathbb{Q}\}$. Then f preserves the distances 1 and $\sqrt{3}$, but it is not an isometry.

Corollary 5.2 *Let X and Y be real Hilbert spaces with* $\dim X > 1$. *If a mapping $f : X \to Y$ preserves the distances* 1 *and* $n\sqrt{3}$ *for some $n \in \mathbb{N}$, then f is an affine isometry.*

Proof

(*a*) Assume that p and q are arbitrary points of X with $\|p - q\| = \sqrt{3}$. Since $\dim X > 1$, we can choose two points p_1 and p_2 of X such that the pair of four points $\{p, p_1, q, p_2\}$ comprises the vertices of a rhombus with unit side length with $\|p_2 - p_1\| = 1$, as we can see in Fig. 5.4.
If we set $x = f(p_1) - f(p)$, $y = f(p_2) - f(p)$, and $z = f(q) - f(p)$, then

$$\|x\| = \|y\| = \|x - y\| = \|z - x\| = \|z - y\| = 1.$$

Hence, we have

$$\|x - y\|^2 = \|x\|^2 - 2\langle x, y\rangle + \|y\|^2 = 1,$$

$$\|z - x\|^2 = \|z\|^2 - 2\langle z, x\rangle + \|x\|^2 = 1,$$

$$\|z - y\|^2 = \|z\|^2 - 2\langle z, y\rangle + \|y\|^2 = 1.$$

It then follows from the last three equalities that

$$\langle x, y\rangle = \frac{1}{2} \quad \text{and} \quad \langle z, x\rangle = \langle z, y\rangle = \frac{1}{2}\|z\|^2.$$

Thus, we obtain

$$\|x + y\|^2 = \|x\|^2 + 2\langle x, y\rangle + \|y\|^2 = 3$$

and

$$\|z\|^2 = \langle z, x\rangle + \langle z, y\rangle = \langle z, x + y\rangle.$$

Moreover, it follows from the Cauchy–Schwarz inequality that

$$\|z\|^2 = \langle z, x + y\rangle = |\langle z, x + y\rangle| \leq \|z\|\|x + y\| = \sqrt{3}\,\|z\|,$$

and hence, $\|z\| = \|f(q) - f(p)\| \leq \sqrt{3}$.

(b) Assume that f preserves the distances 1 and $n\sqrt{3}$ for some integer $n > 1$. Let p and q_1 be any points of X with $\|p - q_1\| = \sqrt{3}$. If we set

$$q_k = p + k(q_1 - p)$$

for all $k \in \{1, 2, \ldots, n\}$, then we obtain $\|q_{k+1} - q_k\| = \|q_1 - p\| = \sqrt{3}$ for any $k \in \{1, 2, \ldots, n-1\}$ and $\|q_n - p\| = n\sqrt{3}$. Since $\dim X > 1$, we can construct n rhombi of unit side length, as shown in Fig. 5.5.

(c) It follows from (a) and (b) that

$$\|f(q_k) - f(q_{k-1})\| \leq \sqrt{3}$$

for any $k \in \{1, 2, \ldots, n\}$, where we set $q_0 = p$, and

$$n\sqrt{3} = \|f(q_n) - f(p)\| \leq \sum_{k=1}^{n} \|f(q_k) - f(q_{k-1})\| \leq n\sqrt{3},$$

since f preserves the distance $n\sqrt{3}$. Thus, we obtain

$$\|f(q_1) - f(p)\| = \|f(q_2) - f(q_1)\| = \cdots = \|f(q_n) - f(q_{n-1})\| = \sqrt{3}.$$

Therefore, the distance $\sqrt{3}$ is also preserved by f. By Theorem 5.1, f is an affine isometry. □

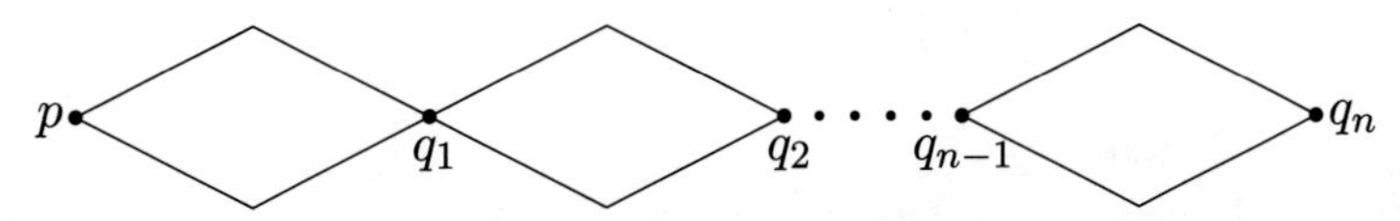

Fig. 5.5 The tail vertex of the preceding rhombus overlaps the head vertex of the succeeding rhombus

5.2 The Case Where 1 and $\sqrt{2}$ Are Preserved

In the previous section, we considered the case where the distances 1 and $\sqrt{3}$ are preserved, as a special case of the Aleksandrov–Rassias problem where the ratio of two distances is not an integer. As in the previous section, we use "points," "elements," and "vectors" interchangeably and treat them as synonyms.

In this section, we will consider the case where the distances 1 and $\sqrt{2}$ are preserved. Xiang proved the following theorem in his papers [71, 72].

Theorem 5.3 (Xiang) *Let X and Y be real Hilbert spaces with* $\dim X > 2$. *If a mapping $f : X \to Y$ preserves the distances* 1 *and* $\sqrt{2}$, *then f is an affine isometry.*

Proof

(*a*) Suppose p_1, p_2, p_3, p_4 are arbitrary points of X that form the vertices of a rhombus whose sides have unit length and whose diagonal lengths are both $\|p_1 - p_3\| = \|p_2 - p_4\| = \sqrt{2}$, as shown in Fig. 5.6.
We claim that the pair of four points $\{f(p_1), f(p_2), f(p_3), f(p_4)\}$ comprises the vertices of a unit rhombus with $\|f(p_1) - f(p_3)\| = \|f(p_2) - f(p_4)\| = \sqrt{2}$.
First, we set $x = f(p_2) - f(p_1)$, $y = f(p_4) - f(p_1)$, and $z = f(p_3) - f(p_1)$. Since f preserves the distances 1 and $\sqrt{2}$, we have

$$\|x\| = \|y\| = \|z - x\| = \|z - y\| = 1 \quad \text{and} \quad \|z\| = \|x - y\| = \sqrt{2}.$$

Moreover, since Y is a real Hilbert space, we obtain

$$\begin{aligned}
\|x - y\|^2 &= \|x\|^2 - 2\langle x, y\rangle + \|y\|^2 = 2, \\
\|z - x\|^2 &= \|z\|^2 - 2\langle z, x\rangle + \|x\|^2 = 1, \\
\|z - y\|^2 &= \|z\|^2 - 2\langle z, y\rangle + \|y\|^2 = 1.
\end{aligned}$$

Thus, it follows that

$$\langle x, y\rangle = 0, \quad \langle z, x\rangle = \langle z, y\rangle = 1,$$

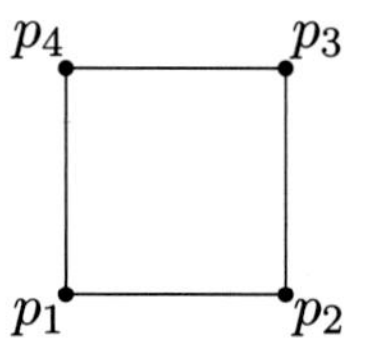

Fig. 5.6 p_1, p_2, p_3, p_4 are the vertices of a rhombus whose side length is 1 and whose diagonals have length $\sqrt{2}$

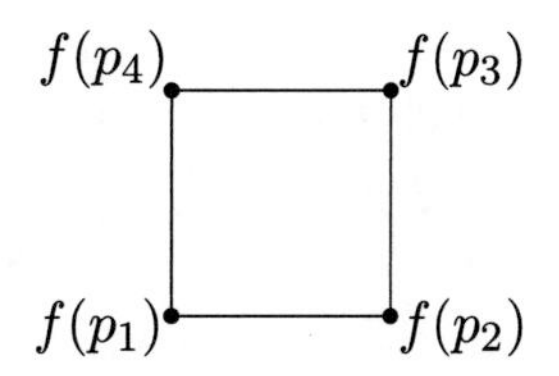

Fig. 5.7 $f(p_1), f(p_2), f(p_3), f(p_4)$ are the vertices of a rhombus whose side length is 1 and whose diagonals have length $\sqrt{2}$

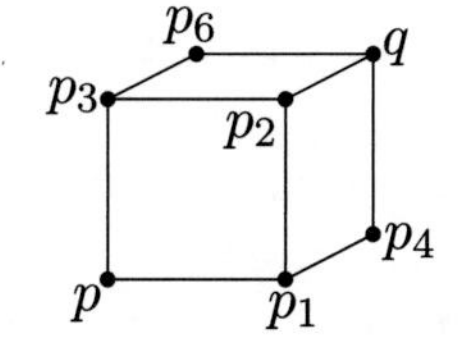

Fig. 5.8 Let p and q be two points whose distance from each other is $\sqrt{3}$. The figure shows a unit cube with the vertices $p, p_1, p_2, p_3, p_4, q, p_6$ and p_5

and

$$\|z - x - y\|^2 = \|z\|^2 - 2\langle z, x\rangle - 2\langle z, y\rangle + \|x\|^2 + 2\langle x, y\rangle + \|y\|^2 = 0.$$

Hence, it follows that $z = x + y$. That is, $f(p_1), f(p_2), f(p_3), f(p_4)$ are the vertices of a unit rhombus with $\|f(p_1) - f(p_3)\| = \|f(p_2) - f(p_4)\| = \sqrt{2}$. In particular, $f(p_3)$ is in the span of $\{f(p_1), f(p_2), f(p_4)\}$. Furthermore, $f(p_3) - f(p_1) = (f(p_2) - f(p_1)) + (f(p_4) - f(p_1))$. (See Fig. 5.7.)

(*b*) Let p and q be any points of X with $\|p - q\| = \sqrt{3}$. Since $\dim X > 2$, we can construct a unit cube, whose vertices are $p, p_1, p_2, p_3, p_4, q, p_6$, and p_5, such that $\|p - q\| = \sqrt{3}$ as we see in Fig. 5.8. (In Fig. 5.8, the point p_5 is not shown.):
It is to be noted that the six pairs of four points $\{p, p_1, p_2, p_3\}$, $\{p_1, p_4, q, p_2\}$, $\{p, p_1, p_4, p_5\}$, $\{p_2, q, p_6, p_3\}$, $\{p, p_5, p_6, p_3\}$, and $\{p_4, q, p_6, p_5\}$ are the sets of vertices of unit rhombi with diagonal lengths $\sqrt{2}$, respectively.

According to (*a*), each of the following six pairs of four points

$$\{f(p), f(p_1), f(p_2), f(p_3)\}, \{f(p_1), f(p_4), f(q), f(p_2)\},$$
$$\{f(p), f(p_1), f(p_4), f(p_5)\}, \{f(p_2), f(q), f(p_6), f(p_3)\},$$
$$\{f(p), f(p_5), f(p_6), f(p_3)\}, \{f(p_4), f(q), f(p_6), f(p_5)\}$$

comprises the vertices of a corresponding unit rhombus with diagonal lengths $\sqrt{2}$. Moreover, $f(p_5)$ is in the span of $\{f(p), f(p_1), f(p_4)\}$ and $f(p_6)$ is in the span of $\{f(p_2), f(q), f(p_3)\}$. Hence, $f(p), f(p_1), f(p_2), f(p_3), f(p_4), f(q), f(p_6), f(p_5)$ are the vertices of a unit cube. Therefore, it is obvious that $\|f(p) - f(q)\| = \sqrt{3}$, which implies that f preserves the distance $\sqrt{3}$. Consequently, due to Theorem 5.1, f is an affine isometry. □

Assume that X and Y are real Hilbert spaces with $\dim X > 2$. If a mapping $f : X \to Y$ transforms every unit square in X into a unit square in Y, it then follows from Theorem 5.3 that f is an affine isometry. This corollary can be proved in a similar way to the proof of Corollary 8.40.

5.3 The Case Where Three Distances Are Preserved

So far we have mainly been concerned with whether every mapping that preserves at most two distances is an isometry. In this section, we will examine under what conditions a mapping that preserves three distances can become an isometry. As we can see in Problem 4.2 (ii), this problem is closely related to the Aleksandrov-Rassias problem.

We introduce a theorem that was proved in a paper [60] by Th. M. Rassias and S. Xiang.

Theorem 5.4 (Rassias and Xiang) *Let X and Y be real normed spaces such that $\dim X > 1$ and Y is strictly convex. Suppose $f : X \to Y$ preserves three distances 1, ρ, and $1 + \rho$, where ρ is any positive constant. Then f is an affine isometry.*

Proof

(a) We claim that f preserves the distance $2 + \rho$. Suppose x and y are arbitrary points of X with $\|x - y\| = 2 + \rho$. If we put

$$x_1 = x + \frac{1}{2+\rho}(y - x) \quad \text{and} \quad x_2 = x + \frac{1+\rho}{2+\rho}(y - x),$$

then we have

$$\begin{gathered}\|x_1 - x\| = 1, \quad \|x_2 - x_1\| = \rho, \quad \|y - x_1\| = 1 + \rho, \\ \|x_2 - x\| = 1 + \rho, \quad \|y - x_2\| = 1.\end{gathered}$$

By our assumption that f preserves the distances 1, ρ, and $1 + \rho$, we obtain

$$\begin{gathered}\|f(x_1) - f(x)\| = 1, \quad \|f(x_2) - f(x_1)\| = \rho, \quad \|f(y) - f(x_1)\| = 1 + \rho, \\ \|f(x_2) - f(x)\| = 1 + \rho, \quad \|f(y) - f(x_2)\| = 1,\end{gathered}$$

and thus, it follows that

$$\begin{aligned}\|f(x_2) - f(x)\| &= 1 + \rho = \|f(x_2) - f(x_1)\| + \|f(x_1) - f(x)\|, \\ \|f(y) - f(x_1)\| &= 1 + \rho = \|f(y) - f(x_2)\| + \|f(x_2) - f(x_1)\|.\end{aligned}$$

Since Y is a real normed space that is strictly convex, we have

$$f(x_1) = \frac{\rho}{1+\rho} f(x) + \frac{1}{1+\rho} f(x_2)$$

and

$$f(x_2) = \frac{\rho}{1+\rho} f(y) + \frac{1}{1+\rho} f(x_1).$$

Thus, $\|f(x) - f(y)\| = 2 + \rho$ for all $x, y \in X$ with $\|x - y\| = 2 + \rho$. Therefore, f preserves the distance $2 + \rho$.

(b) We claim that f preserves the distance $2 + 2\rho$. Let x and y be arbitrary points of X with $\|x - y\| = 2 + 2\rho$. If we set

$$x_1 = x + \frac{1+\rho}{2+2\rho}(y - x) \quad \text{and} \quad x_2 = x + \frac{2+\rho}{2+2\rho}(y - x),$$

then

$$\|x_1 - x\| = 1 + \rho, \quad \|x_2 - x_1\| = 1, \quad \|y - x_1\| = 1 + \rho,$$
$$\|x_2 - x\| = 2 + \rho, \quad \|y - x_2\| = \rho.$$

Since f preserves the distances 1, ρ, $1 + \rho$, and $2 + \rho$ by (a), we have

$$\|f(x_1) - f(x)\| = 1 + \rho, \quad \|f(x_2) - f(x_1)\| = 1, \quad \|f(y) - f(x_1)\| = 1 + \rho,$$
$$\|f(x_2) - f(x)\| = 2 + \rho, \quad \|f(y) - f(x_2)\| = \rho.$$

Hence, it follows that

$$\|f(x_2) - f(x)\| = 2 + \rho = \|f(x_2) - f(x_1)\| + \|f(x_1) - f(x)\|,$$
$$\|f(y) - f(x_1)\| = 1 + \rho = \|f(y) - f(x_2)\| + \|f(x_2) - f(x_1)\|.$$

Since Y is a real normed space that is strictly convex, it holds that

$$f(x_1) = \frac{1}{2+\rho} f(x) + \frac{1+\rho}{2+\rho} f(x_2)$$

and

$$f(x_2) = \frac{1}{1+\rho} f(y) + \frac{\rho}{1+\rho} f(x_1).$$

Hence, we get

$$f(x) = (2+\rho)f(x_1) - (1+\rho)f(x_2) \quad \text{and} \quad f(y) = (1+\rho)f(x_2) - \rho f(x_1).$$

Thus, $\|f(x) - f(y)\| = 2 + 2\rho$ for all $x, y \in X$ with $\|x - y\| = 2 + 2\rho$. Therefore, f preserves the distance $2 + 2\rho$.

(c) Consequently, by (a), (b), and the Benz-Berens theorem, we may conclude that f is an affine isometry. □

Remark 5.5 Theorem 5.4 is also true if the distances 1, ρ and $1 + \rho$ are replaced by σ, τ and $\sigma + \tau$, where σ and τ are positive constants. In particular, if the ratio of the two positive constants σ and τ is an integer, then f is an affine isometry due to the Benz-Berens theorem.

Corollary 5.6 *Let X and Y be real normed spaces such that* $\dim X > 1$ *and Y is strictly convex. Assume that* ρ *is a real constant greater than* 1. *If a mapping* $f : X \to Y$ *preserves three distances* 1, ρ, *and* $\{\rho\}$, *where* $\{\rho\} = \rho - [\rho]$ *and* $[\rho]$ *is the largest integer not exceeding* ρ, *then* f *is an affine isometry.*

Proof

(a) If $1 < \rho < 2$, then $\rho = 1 + \{\rho\}$, then it follows from Theorem 5.4 that f is an affine isometry.

(b) When $\rho = 2$, it then follows from Theorem 2.5 and theorem of Benz and Berens that f is an affine isometry.

(c) Assume that $\rho > 2$ and $[\rho] = n > 1$. Let x and y be arbitrary points of X with $\|x - y\| = n$. If we set

$$z = x + \frac{\rho}{n}(y - x) \quad \text{and} \quad x_k = x + \frac{k}{n}(y - x)$$

for any $k \in \{0, 1, \ldots, n\}$, then $\|x_k - x_{k+1}\| = 1$ for each $k \in \{0, 1, \ldots, n-1\}$, where $x_0 = x$ and $x_n = y$. Moreover, we have $\|x - z\| = \rho$ and $\|y - z\| = \{\rho\}$. Since f preserves the distances 1, ρ, and $\{\rho\}$, we obtain

$$\begin{aligned}
\rho &= \|f(x) - f(z)\| \\
&\leq \|f(x) - f(x_n)\| + \|f(y) - f(z)\| \\
&\leq \sum_{k=0}^{n-1} \|f(x_k) - f(x_{k+1})\| + \|f(y) - f(z)\| \\
&= [\rho] + \{\rho\} \\
&= \rho.
\end{aligned}$$

Hence, it follows that

$$\|f(x) - f(y)\| = \|f(x) - f(x_n)\| = \sum_{k=0}^{n-1} \|f(x_k) - f(x_{k+1})\| = n.$$

Therefore, f preserves the distance $[\rho] = n > 1$. Consequently, it follows from the Benz–Berens theorem that f is an affine isometry. □

The following theorem proved by Xiang generalizes Theorem 5.1 (see [72, Theorem 2.3]).

Theorem 5.7 (Xiang) *Let X and Y be real Hilbert spaces with $\dim X > 1$. Assume that ρ and σ are positive constants that satisfy $0 < \rho \leq 2\sigma$. If a mapping $f : X \to Y$ preserves the distances ρ, σ, and $\sqrt{2\rho^2 + \sigma^2}$, then f is an affine isometry.*

Proof

(*a*) If $\rho = 2\sigma$, then f is an affine isometry due to Theorem 2.5 and the Benz–Berens theorem.

(*b*) Now we assume that $0 < \rho < 2\sigma$. Let $\{p_1, p_2, p_3, p_4\}$ be an arbitrary pair of four points that comprises the vertices of a parallelogram in X with $\|p_4 - p_1\| = \|p_3 - p_2\| = \rho$, $\|p_3 - p_4\| = \|p_2 - p_1\| = \sigma$, $\|p_1 - p_3\| = \sqrt{2\rho^2 + \sigma^2}$, and $\|p_2 - p_4\| = \sigma$, as we see in Fig. 5.9.
Then, the pair of four points $\{f(p_1), f(p_2), f(p_3), f(p_4)\}$ comprises the vertices of a parallelogram lying in Y with $\|f(p_4) - f(p_1)\| = \|f(p_3) - f(p_2)\| = \rho$, $\|f(p_3) - f(p_4)\| = \|f(p_2) - f(p_1)\| = \sigma$, $\|f(p_1) - f(p_3)\| = \sqrt{2\rho^2 + \sigma^2}$, $\|f(p_2) - f(p_4)\| = \sigma$, and $f(p_3) - f(p_1) = (f(p_4) - f(p_1)) + (f(p_2) - f(p_1))$.

(*c*) Let p and q be arbitrary points of X with $\|p - q\| = 2\rho$ and let $p_1 = \frac{1}{2}(p + q)$. Since $\dim X > 1$, there are points p_2 and p_3 in X such that $\{p, p_1, p_2, p_3\}$ and $\{p_1, p_3, p_2, q\}$ are the sets of vertices of two parallelograms with $\|p_1 - p\| = \rho$ and $\|p_2 - p_1\| = \|p_3 - p_1\| = \sigma$. (See Fig. 5.10.)
According to (*b*), $\{f(p), f(p_1), f(p_2), f(p_3)\}$ and $\{f(p_1), f(p_3), f(p_2), f(q)\}$ are the sets of vertices of two parallelograms with side lengths ρ and σ, respectively.

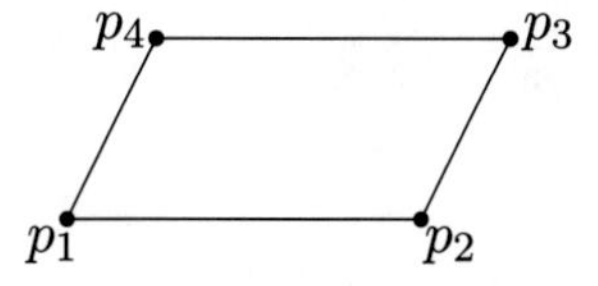

Fig. 5.9 The four points p_1, p_2, p_3, p_4 are vertices of a parallelogram, where $\|p_4 - p_1\| = \|p_3 - p_2\| = \rho$, $\|p_3 - p_4\| = \|p_2 - p_1\| = \|p_2 - p_4\| = \sigma$, and $\|p_1 - p_3\| = \sqrt{2\rho^2 + \sigma^2}$ for $0 < \rho < 2\sigma$

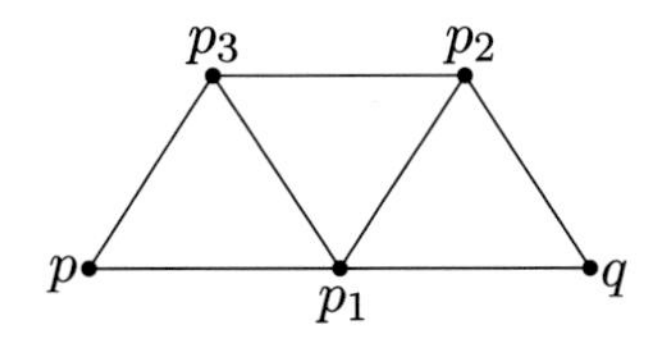

Fig. 5.10 Let p and q be two points separated by 2ρ and the midpoint of these points be p_1. Two more points p_2 and p_3 are chosen such that p, p_1, p_2, p_3 and p_1, q, p_2, p_3 are the vertices of two parallelograms with side lengths ρ and σ

(*d*) If we set $x = f(p_1) - f(p)$ and $y = f(p_3) - f(p)$, then $x - y = f(p_1) - f(p_3)$, $x + y = f(p_2) - f(p)$, and $f(q) - f(p_3) = (f(p_1) - f(p_3)) + (f(p_2) - f(p_3))$. Thus, the points $f(p), f(p_1), f(p_2), f(p_3)$, and $f(q)$ lie on a plane in Y. Hence, we have $f(q) - f(p) = 2x = 2(f(p_1) - f(p))$ and $\|f(q) - f(p)\| = 2\rho$. Therefore, f preserves the distance 2ρ. Consequently, by Theorem 2.5 and the Benz–Berens theorem, we may conclude that f is an affine isometry. □

We can easily prove the following corollary by exchanging the roles of ρ and σ in Theorem 5.7.

Corollary 5.8 *Let X and Y be real Hilbert spaces with* $\dim X > 1$. *Assume that ρ and σ are positive constants that satisfy* $0 < \sigma \leq 2\rho$. *If a mapping* $f : X \to Y$ *preserves the distances* ρ, σ, *and* $\sqrt{\rho^2 + 2\sigma^2}$, *then f is an affine isometry.*

Let X and Y be real Hilbert spaces with $\dim X > 1$. Assume that a mapping $f : X \to Y$ preserves the two distances ρ and σ. Suppose $\{p_1, p_2, p_3, p_4\}$ is a pair of four points that comprise the vertices of a parallelogram in X with $\|p_2 - p_1\| = \|p_3 - p_4\| = \rho$ and $\|p_3 - p_2\| = \|p_4 - p_1\| = \|p_2 - p_4\| = \sigma$. By the same way presented in the proof of Corollary 5.2, we can obtain $\|f(p_3) - f(p_1)\| \leq \sqrt{2\rho^2 + \sigma^2}$. If f preserves the distance $n\sqrt{2\rho^2 + \sigma^2}$, using a similar method as the proof of Corollary 5.2, we can show that f preserves the distance $\sqrt{2\rho^2 + \sigma^2}$.

Corollary 5.9 *Let X and Y be real Hilbert spaces with* $\dim X > 1$. *Assume that ρ and σ are positive constants that satisfy* $0 < \rho \leq 2\sigma$. *If a mapping* $f : X \to Y$ *preserves the distances* ρ, σ, *and* $n\sqrt{2\rho^2 + \sigma^2}$ *for some* $n \in \mathbb{N}$, *then f is an affine isometry.*

To prove the following corollary, we only need to exchange the roles of ρ and σ in Corollary 5.9.

Corollary 5.10 *Let X and Y be real Hilbert spaces with* $\dim X > 1$. *Assume that ρ and σ are positive constants that satisfy* $0 < \sigma \leq 2\rho$. *If a mapping* $f : X \to Y$ *preserves the distances* ρ, σ, *and* $n\sqrt{\rho^2 + 2\sigma^2}$ *for some* $n \in \mathbb{N}$, *then f is an affine isometry.*

In addition, Xiang proved the following theorem in terms of three distance-preserving mappings.

Theorem 5.11 (Xiang) *Let X and Y be real Hilbert spaces with* $\dim X > 1$. *Assume that ρ is a real number with $0 \leq \rho \leq 2$. If a mapping $f : X \to Y$ preserves the distances 1, ρ, and $n\sqrt{4-\rho^2}$ for some integer $n > 1$, then f is an affine isometry.*

Proof

(*a*) If $\rho = 0$ or $\rho = 2$, it then follows from Theorem 2.5 and Benz–Berens theorem that f is an affine isometry.

(*b*) We assume that $0 < \rho < 2$. Suppose p and q are arbitrary points in X with $\|p-q\| = \sqrt{4-\rho^2}$. We assert that $\|f(p)-f(q)\| \leq \sqrt{4-\rho^2}$. Since $\dim X > 1$, we can select two points p_1 and p_2 in X such that $\{p, p_1, q, p_2\}$ comprises the vertices of some rhombus with $\|p_1-p\| = \|p_2-p\| = \|q-p_1\| = \|q-p_2\| = 1$ and $\|p_2-p_1\| = \rho$, as shown in Fig. 5.11.
Since f preserves 1 and ρ, it follows that $\|f(p_1)-f(p)\| = \|f(p_2)-f(p)\| = \|f(q)-f(p_1)\| = \|f(q)-f(p_2)\| = 1$ and $\|f(p_2)-f(p_1)\| = \rho$.
We set $x = f(p_1)-f(p)$, $y = f(p_2)-f(p)$, and $z = f(q)-f(p)$. Then we obtain

$$\|x-y\|^2 = \|x\|^2 - 2\langle x, y\rangle + \|y\|^2 = \rho^2,$$
$$\|z-x\|^2 = \|z\|^2 - 2\langle z, x\rangle + \|x\|^2 = 1,$$
$$\|z-y\|^2 = \|z\|^2 - 2\langle z, y\rangle + \|y\|^2 = 1.$$

Hence, we have

$$\|x+y\|^2 = \|x\|^2 + 2\langle x, y\rangle + \|y\|^2 = 4-\rho^2 \quad \text{and} \quad \langle z, x\rangle = \langle z, y\rangle = \frac{1}{2}\|z\|^2.$$

Thus, it follows from the Cauchy–Schwarz inequality that

$$\|z\|^2 = \langle z, x+y\rangle \leq |\langle z, x+y\rangle| \leq \|z\|\|x+y\| = \sqrt{4-\rho^2}\|z\|.$$

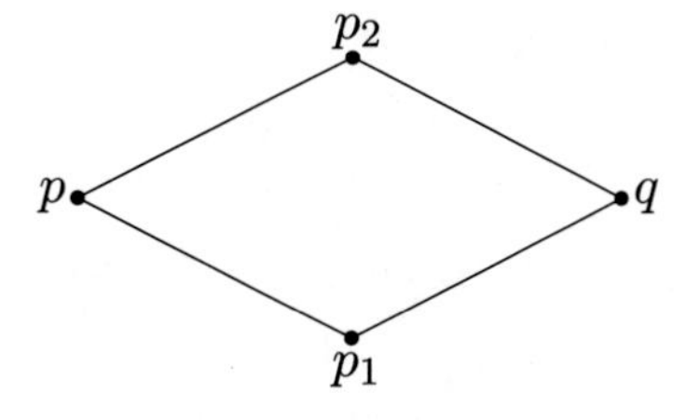

Fig. 5.11 The four points p, p_1, q, p_2 are the vertices of a rhombus with side length 1, $\|p-q\| = \sqrt{4-\rho^2}$ and with $\|p_2-p_1\| = \rho$ for $0 < \rho < 2$

Therefore, we obtain

$$\|f(q) - f(p)\| = \|z\| \le \sqrt{4 - \rho^2}.$$

(c) Assume that f preserves the distances 1 and $n\sqrt{4-\rho^2}$ for some integer $n > 1$. Let p and q_1 be points of X with $\|q_1 - p\| = \sqrt{4-\rho^2}$. If we set

$$q_k = p + k(q_1 - p)$$

for every $k \in \{1, 2, \ldots, n\}$, then $\|q_{k+1} - q_k\| = \|q_1 - p\| = \sqrt{4-\rho^2}$ for all $k \in \{1, 2, \ldots, n-1\}$ and $\|q_n - p\| = n\sqrt{4-\rho^2}$. Since $\dim X > 1$, we can construct n rhombi of unit side length, as shown in Fig. 5.12.

Now, it follows from (b) that

$$\|f(q_k) - f(q_{k-1})\| \le \sqrt{4-\rho^2}$$

for all $k \in \{1, 2, \ldots, n\}$, where we set $q_0 = p$. Thus, we have

$$\|f(q_n) - f(p)\| \le \sum_{k=1}^{n} \|f(q_k) - f(q_{k-1})\| \le n\sqrt{4-\rho^2}.$$

Since Y is a real Hilbert space and f preserves the distance $n\sqrt{4-\rho^2}$, we obtain

$$\|f(q_k) - f(q_{k-1})\| = \sqrt{4-\rho^2}$$

for all $k \in \{1, 2, \ldots, n\}$. Therefore, f preserves the distance $\sqrt{4-\rho^2}$. Since we assumed that f preserves the distance $n\sqrt{4-\rho^2}$ for some integer $n > 1$, it follows from Theorem 2.5 and the Benz–Berens theorem that f is an affine isometry. □

If we set $\rho = n\sqrt{4-\rho^2} = \frac{2n}{\sqrt{n^2+1}}$ for some integer $n > 1$, then $0 \le \rho \le 2$. The following corollary is an immediate consequence of Theorem 5.11.

Corollary 5.12 *Let X and Y be real Hilbert spaces with* $\dim X > 1$. *If a mapping $f : X \to Y$ preserves the distances* 1 *and* $\frac{2n}{\sqrt{n^2+1}}$ *for some integer $n > 1$, then f is an affine isometry.*

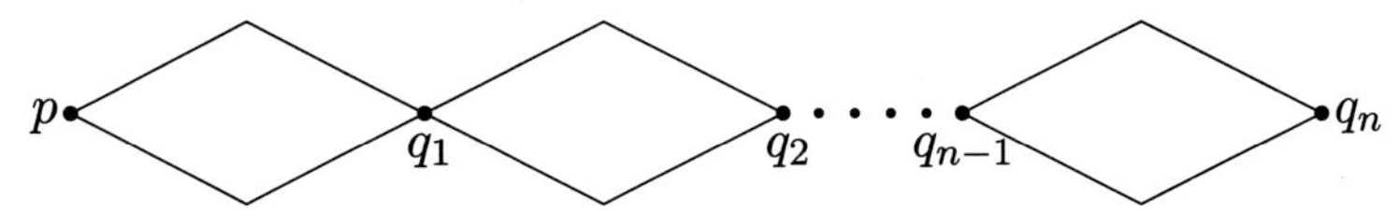

Fig. 5.12 The tail vertex of the preceding rhombus and the head vertex of the succeeding rhombus are the same

Inequalities for Distances Between Points 6

Abstract

In this chapter, we will generalize the parallelogram law to all cases with a given number of points. In Sect. 6.1, we prove the short diagonals lemma, which is a generalization of the parallelogram law for the given four points in the inner product space. Moreover, applying the short diagonals lemma and the parallelogram law, we prove an inequality for the distances between any two points among the given six points. Section 6.2 is devoted to a generalization of the short diagonals lemma to an inequality for the distances among the given $2n$ points in the inner product space. In this section, we conceive and prove an inequality involving all distances between an even number of points. It seems to be more difficult to prove an inequality for the distances between any two points among an odd number of points than among an even number of points. In Sect. 6.3, we use a new method other than the short diagonals lemma to prove an inequality for the distances between any two of the given five points. It is somewhat surprising that the golden ratio appears in this inequality. In Sect. 6.4, an inequality is introduced that describes in general the relationship between distances among the given n points. When studying inequalities for distances between given n points, it is recommended to pay attention to the following two requirements:

(†) All possible distances between points are included in the inequality.
(‡) The necessary and sufficient condition for the equality sign is presented.

The main results presented in this chapter have been extracted from the papers by Jung (Nonlinear Anal 62(4):675–681, 2005), Jung and Lee (J Math Anal Appl 324(2):1363–1369, 2006), Jung and Nam (J Math Inequal 12(4):1189–1199, 2018), Jung and Nam (J Math Inequal 13(4):969–981, 2019) and explained in detail so that the reader can easily understand them.

S.-M. Jung, *Aleksandrov-Rassias Problems on Distance Preserving Mappings*,
Frontiers in Mathematics, https://doi.org/10.1007/978-3-031-77613-7_6

6.1 An Inequality for Distances Among Six Points

Throughout this chapter, unless otherwise stated, let H be a real (or complex) inner product space, $\langle\cdot,\cdot\rangle$ the inner product for H, and let $\|\cdot\|$ be the norm on H defined by $\|x\| = \sqrt{\langle x, x\rangle}$ for all $x \in H$. In this book, we often use the terms "points," "elements," and "vectors" interchangeably and treat them as synonyms.

If $\dim H > 1$, then the *Pythagorean theorem* states that for each pair of three points x, y, z of H, the equality

$$\|x - z\|^2 = \|y - x\|^2 + \|z - y\|^2$$

holds if and only if the vectors $y - x$ and $z - y$ are orthogonal to each other. This is one of the most important theorems in mathematics.

In connection with this topic, we first consider the case of four points x, y, z, w of H. The parallelogram law states that the equality

$$\|y - x\|^2 + \|z - y\|^2 + \|w - z\|^2 + \|x - w\|^2 = \|z - x\|^2 + \|w - y\|^2$$

holds if and only if $y - x$, $z - y$, $w - z$, $x - w$ are the sides of a (possibly degenerate) parallelogram with diagonals $z - x$ and $w - y$ (see Theorem 1.42). However, if $y - x$, $z - y$, $w - z$, $x - w$ are not the sides of a (possibly degenerate) parallelogram with diagonals $z - x$ and $w - y$, the following strict inequality holds (cf. [49, Lemma 15.4.2]):

$$\|y - x\|^2 + \|z - y\|^2 + \|w - z\|^2 + \|x - w\|^2 > \|z - x\|^2 + \|w - y\|^2.$$

The parallelogram law can be generalized to an inequality that can be applied to any four points in an inner product space. This inequality is known as the *short diagonals lemma* and was proved in [49, Lemma 15.4.2] for the case of Euclidean spaces.

Because of its pedagogical importance, we will first introduce the proof of the short diagonals lemma for the case of inner product spaces.

Lemma 6.1 (Short Diagonals Lemma) *If H is a real (or complex) inner product space, then*

$$\|y - x\|^2 + \|z - y\|^2 + \|w - z\|^2 + \|x - w\|^2 \geq \|z - x\|^2 + \|w - y\|^2$$

for all $x, y, z, w \in H$.

Proof We will prove the lemma when H is a complex inner product space. Let x, y, z, w be arbitrary vectors of H. Then, we get

$$
\begin{aligned}
&\|y-x\|^2+\|z-y\|^2+\|w-z\|^2+\|x-w\|^2-\|z-x\|^2-\|w-y\|^2\\
&\quad=\langle y-x,y-x\rangle+\langle z-y,z-y\rangle+\langle w-z,w-z\rangle+\langle x-w,x-w\rangle\\
&\qquad-\langle z-x,z-x\rangle-\langle w-y,w-y\rangle\\
&\quad=\langle y-x,y-x\rangle+\langle z,z\rangle-\langle y,z\rangle-\overline{\langle y,z\rangle}+\langle y,y\rangle\\
&\qquad+\langle w-z,w-z\rangle+\langle x,x\rangle-\langle w,x\rangle-\overline{\langle w,x\rangle}+\langle w,w\rangle-\langle z,z\rangle\\
&\qquad+\langle x,z\rangle+\overline{\langle x,z\rangle}-\langle x,x\rangle-\langle w,w\rangle+\langle y,w\rangle+\overline{\langle y,w\rangle}-\langle y,y\rangle\\
&\quad=\langle y-x,y-x\rangle+\langle w-z,w-z\rangle+\langle y,w-z\rangle+\overline{\langle y,w-z\rangle}\\
&\qquad+\langle z-w,x\rangle+\overline{\langle z-w,x\rangle}\\
&\quad=\langle y-x,y-x\rangle+\langle w-z,w-z\rangle+\langle y-x,w-z\rangle+\overline{\langle y-x,w-z\rangle}\\
&\quad=\langle y-x+w-z,y-x+w-z\rangle\\
&\quad\geq 0,
\end{aligned}
$$

which completes the proof. □

By applying the short diagonals lemma and the parallelogram law, we prove an inequality for the distances between any two points among the given six points. The following theorem proved by S.-M. Jung [27] may be regarded as a generalization of the short diagonals lemma.

Theorem 6.2 (Jung) *Let H be a real (or complex) inner product space.*

(*i*) *The inequality*

$$
\begin{aligned}
&2\big(\|z-u\|^2+\|x-v\|^2+\|y-w\|^2\big)\\
&\quad\leq\|v-u\|^2+\|z-v\|^2+\|x-z\|^2+\|u-x\|^2\\
&\qquad+\|w-u\|^2+\|z-w\|^2+\|y-z\|^2+\|u-y\|^2\\
&\qquad+\|w-v\|^2+\|x-w\|^2+\|y-x\|^2+\|v-y\|^2
\end{aligned}
\tag{6.1}
$$

holds for any six points $u, v, w, x, y, z \in H$.

(*ii*) *The equality sign is true if and only if each pair of four points $\{u, v, z, x\}$, $\{u, w, z, y\}$, $\{v, w, x, y\}$ comprises the vertices of an appropriate (possibly degenerate) parallelogram such that $z-u$ and $x-v$, resp., $z-u$ and $y-w$, resp., $x-v$ and $y-w$ are the diagonals of the corresponding parallelogram.*

Proof

(i) Applying Lemma 6.1 to each ordered quadruplet, (u, v, z, x), (u, w, z, y), (v, w, x, y) consecutively, we have

$$\begin{aligned}
&\|z-u\|^2+\|x-v\|^2 \le \|v-u\|^2+\|z-v\|^2+\|x-z\|^2+\|u-x\|^2,\\
&\|z-u\|^2+\|y-w\|^2 \le \|w-u\|^2+\|z-w\|^2+\|y-z\|^2+\|u-y\|^2, \quad (6.2)\\
&\|x-v\|^2+\|y-w\|^2 \le \|w-v\|^2+\|x-w\|^2+\|y-x\|^2+\|v-y\|^2,
\end{aligned}$$

respectively. By summing up the three inequalities, we easily obtain the desired inequality (6.1).

(ii) We now assume that the equality sign holds in (6.1). Lemma 6.1 implies that every inequality in (6.2) is true. Hence, it is easy to check that strict inequality in at least one of three inequalities of (6.2) implies the strict inequality sign in (6.1), a contradiction. Hence, we conclude that if the equality sign holds in (6.1), then the equality sign has to be true in each inequality of (6.2). That is, in view of parallelogram law, we conclude that each of $\{v-u, z-v, x-z, u-x\}$, $\{w-u, z-w, y-z, u-y\}$, and $\{w-v, x-w, y-x, v-y\}$ comprises the sides of an appropriate (possibly degenerate) parallelogram such that $z-u$ and $x-v$, resp., $z-u$ and $y-w$, resp., $x-v$ and $y-w$ are the diagonals of the corresponding parallelogram.

Conversely, we assume that each of $\{v-u, z-v, x-z, u-x\}$, $\{w-u, z-w, y-z, u-y\}$, and $\{w-v, x-w, y-x, v-y\}$ comprises the sides of an appropriate (possibly degenerate) parallelogram such that $z-u$ and $x-v$, resp., $z-u$ and $y-w$, resp., $x-v$ and $y-w$ are the diagonals of the corresponding parallelogram. Due to the parallelogram law, the equality sign holds in each inequality of (6.2), and hence, the equality sign holds in (6.1). □

Remark 6.3 Let H be a real (or complex) inner product space.

(i) Every possible distance between two points among six different points is included in the inequality (6.1).

(ii) Theorem 6.2 (ii) provides a necessary and sufficient condition to ensure the equality sign in the inequality (6.1).

(iii) We ask whether there is another inequality besides the inequality (6.1) that satisfies the aforementioned conditions (i) and (ii).

6.2 An Inequality for Distances Among $2n$ Points

In this section, using the short diagonals lemma, we will generalize the inequality (6.1) to an inequality for the distances among the given $2n$ points, where n is an integer greater than 1.

The following theorem was presented in the paper [36] by S.-M. Jung and K.-S. Lee.

Theorem 6.4 (Jung and Lee) *Let n be an integer greater than 1, let H be a real (or complex) inner product space, and let p_{ik} be any distinct $2n$ points of H, where $i \in \{1, 2, \ldots, n\}$ and $k \in \{1, 2\}$.*

(*i*) *It holds that*

$$\sum_{\substack{i, j \in \{1, 2, \ldots, n\} \\ k, \ell \in \{1, 2\} \\ i < j}} \|p_{ik} - p_{j\ell}\|^2 \geq (n-1) \sum_{i \in \{1,2,\ldots,n\}} \|p_{i1} - p_{i2}\|^2. \tag{6.3}$$

(*ii*) *The equality sign holds in the inequality (6.3) if and only if for all $i, j \in \{1, 2, \ldots, n\}$ with $i < j$, the pair of four points $\{p_{i1}, p_{i2}, p_{j1}, p_{j2}\}$ comprises the vertices of an appropriate (possibly degenerate) parallelogram such that p_{i1} and p_{j1} are the opposite vertices to p_{i2} and p_{j2}, respectively.*

Proof

(*i*) We will prove the inequality (6.3) by mathematical induction. For $n = 2$, inequality (6.3) is clearly correct according to Lemma 6.1. Now we assume that the inequality (6.3) holds for any $2n$ points p_{ik}, $i \in \{1, 2, \ldots, n\}$ and $k \in \{1, 2\}$, where n is some integer greater than 1.

Assume that p_{ik} are $2(n+1)$ arbitrary points, where $i \in \{1, 2, \ldots, n+1\}$ and $k \in \{1, 2\}$. According to our induction assumption, the inequality (6.3) holds for the $2n$ points p_{ik}, where $i \in \{1, 2, \ldots, n\}$ and $k \in \{1, 2\}$. If we choose four points p_{i1}, $p_{(n+1)1}$, p_{i2}, and $p_{(n+1)2}$, then Lemma 6.1 implies that

$$\sum_{k,\ell \in \{1,2\}} \|p_{ik} - p_{(n+1)\ell}\|^2 \geq \|p_{i1} - p_{i2}\|^2 + \|p_{(n+1)1} - p_{(n+1)2}\|^2 \tag{6.4}$$

for each $i \in \{1, 2, \ldots, n\}$.

By summing the inequalities in (6.3) and (6.4) for all $i \in \{1, 2, \ldots, n\}$, we obtain

$$\sum_{\substack{i,j\in\{1,2,\ldots,n\}\\ k,\ell\in\{1,2\}\\ i<j}} \|p_{ik}-p_{j\ell}\|^2 + \sum_{\substack{i\in\{1,2,\ldots,n\}\\ k,\ell\in\{1,2\}}} \|p_{ik}-p_{(n+1)\ell}\|^2$$

$$\geq (n-1)\sum_{i\in\{1,2,\ldots,n\}} \|p_{i1}-p_{i2}\|^2$$

$$+\sum_{i\in\{1,2,\ldots,n\}} \left(\|p_{i1}-p_{i2}\|^2+\|p_{(n+1)1}-p_{(n+1)2}\|^2\right)$$

and hence,

$$\sum_{\substack{i,j\in\{1,2,\ldots,n+1\}\\ k,\ell\in\{1,2\}\\ i<j}} \|p_{ik}-p_{j\ell}\|^2 \geq n \sum_{i\in\{1,2,\ldots,n+1\}} \|p_{i1}-p_{i2}\|^2,$$

which is the inequality (6.3) for $n+1$. Finally, it follows from the induction conclusion that (6.3) holds for all integers $n>1$.

(ii) According to Lemma 6.1, the inequality

$$\sum_{k,\ell\in\{1,2\}} \|p_{ik}-p_{j\ell}\|^2 \geq \|p_{i1}-p_{i2}\|^2+\|p_{j1}-p_{j2}\|^2 \tag{6.5}$$

holds for any $i,j\in\{1,2,\ldots,n\}$ with $i<j$. Furthermore, it is obvious that

$$\sum_{\substack{i,j\in\{1,2,\ldots,n\}\\ i<j}} \left(\|p_{i1}-p_{i2}\|^2+\|p_{j1}-p_{j2}\|^2\right) = (n-1)\sum_{i\in\{1,2,\ldots,n\}} \|p_{i1}-p_{i2}\|^2. \tag{6.6}$$

If we assume that the equality sign holds in (6.3) and the strict inequality holds in (6.5) for some $i,j\in\{1,2,\ldots,n\}$ with $i<j$, then it follows from (6.5) and (6.6) that

$$\sum_{\substack{i,j\in\{1,2,\ldots,n\}\\ k,\ell\in\{1,2\}\\ i<j}} \|p_{ik}-p_{j\ell}\|^2 = \sum_{\substack{i,j\in\{1,2,\ldots,n\}\\ i<j}} \sum_{k,\ell\in\{1,2\}} \|p_{ik}-p_{j\ell}\|^2$$

$$> \sum_{\substack{i,j\in\{1,2,\ldots,n\}\\ i<j}} \left(\|p_{i1}-p_{i2}\|^2+\|p_{j1}-p_{j2}\|^2\right)$$

$$= (n-1)\sum_{i\in\{1,2,\ldots,n\}} \|p_{i1}-p_{i2}\|^2,$$

which contradicts our assumption that the equality sign holds in (6.3). Hence, equality sign in (6.3) implies the equality sign in (6.5) for $i, j \in \{1, 2, \ldots, n\}$ with $i < j$. That means, according to the parallelogram law: If the equality sign in (6.3) holds, then for any $i, j \in \{1, 2, \ldots, n\}$ with $i < j$, the pair of four points $\{p_{i1}, p_{i2}, p_{j1}, p_{j2}\}$ comprises the vertices of a (possibly degenerate) parallelogram such that p_{i1} and p_{j1} are the opposite vertices to p_{i2} and p_{j2}, respectively.

Conversely, assume that for any $i, j \in \{1, 2, \ldots, n\}$ with $i < j$, the pair of four points $\{p_{i1}, p_{i2}, p_{j1}, p_{j2}\}$ comprises the vertices of an appropriate (possibly degenerate) parallelogram such that p_{i1} and p_{j1} are the opposite vertices to p_{i2} and p_{j2}, respectively. It then follows from the parallelogram law that the equality sign holds in (6.5) for all $i, j \in \{1, 2, \ldots, n\}$ with $i < j$. Hence, it follows from (6.5) and (6.6) that

$$\begin{aligned}
\sum_{\substack{i, j \in \{1,2,\ldots,n\} \\ k, \ell \in \{1,2\} \\ i<j}} \|p_{ik} - p_{j\ell}\|^2 &= \sum_{\substack{i, j \in \{1,2,\ldots,n\} \\ i<j}} \sum_{k,\ell \in \{1,2\}} \|p_{ik} - p_{j\ell}\|^2 \\
&= \sum_{\substack{i, j \in \{1,2,\ldots,n\} \\ i<j}} \left(\|p_{i1} - p_{i2}\|^2 + \|p_{j1} - p_{j2}\|^2 \right) \\
&= (n-1) \sum_{i \in \{1,2,\ldots,n\}} \|p_{i1} - p_{i2}\|^2,
\end{aligned}$$

which confirms that the equality sign in (6.3) holds. □

Theorem 6.4 was already proved in [27, Lemma 1] and [27, Theorem 2] for $n = 2$ and $n = 3$, respectively. Indeed, Theorem 6.4 becomes the short diagonals lemma when $n = 2$ and becomes Theorem 6.2 when $n = 3$.

Remark 6.5 Let H be a real (or complex) inner product space and $n > 1$.

- (i) All possible distances between two points among the given $2n$ points of H are included in the inequality (6.3).
- (ii) Theorem 6.4 (ii) provides a necessary and sufficient condition to ensure the equality sign in inequality (6.3).
- (iii) Besides the inequality (6.3), is there another inequality that satisfies the two conditions (i) and (ii) aforementioned?

6.3 An Inequality for Distances Among Five Points

In the previous sections, using the short diagonals lemma, we proved the inequality for the distances between any two of the $2n$ points, where n is an integer greater than 1. It seems more difficult to prove an inequality for the distances between any two points among an odd number of points than among an even number of points.

Recently, S.-M. Jung and D. Nam [37] succeeded in proving an inequality for the distances between any two among the five given points.

We denote by ϕ the *golden ratio*, i.e., $\phi = \frac{1+\sqrt{5}}{2}$. Then $\phi^2 - \phi - 1 = 0$, and ϕ is the ratio of a diagonal to a side in a regular pentagon. It is somewhat surprising that the golden ratio appears in an inequality for the distances between every two points among five points.

Theorem 6.6 (Jung and Nam) *Let H be a real (or complex) inner product space.*

(*i*) *For any five points $x_1, x_2, x_3, x_4, x_5 \in H$, the following inequality holds:*

$$\begin{aligned} &\phi^2\big(\|x_1 - x_2\|^2 + \|x_2 - x_3\|^2 + \|x_3 - x_4\|^2 \\ &\quad + \|x_4 - x_5\|^2 + \|x_5 - x_1\|^2\big) \\ &\geq \|x_1 - x_3\|^2 + \|x_2 - x_4\|^2 + \|x_3 - x_5\|^2 \\ &\quad + \|x_4 - x_1\|^2 + \|x_5 - x_2\|^2. \end{aligned} \tag{6.7}$$

(*ii*) *The equality sign holds if and only if*

$$x_4 = x_1 - \phi x_2 + \phi x_3 \quad \text{and} \quad x_5 = \phi x_1 - \phi x_2 + x_3 \tag{6.8}$$

for any $x_1, x_2, x_3 \in H$.

Proof We will only prove this theorem if H is a complex inner product space.

(*i*) For notational convenience, we set $x_6 = x_1$, $x_7 = x_2$, and $x_8 = x_3$. We set

$$S_j = \sum_{i=1}^{5} \langle x_i, x_{i+j} \rangle$$

for each $j \in \{0, 1, 2\}$. Then, for any $j \in \{0, 1, 2\}$, we obtain

$$\begin{aligned}
&\sum_{i=1}^{5} \|x_i - x_{i+j}\|^2 \\
&= \sum_{i=1}^{5} \langle x_i - x_{i+j},\, x_i - x_{i+j}\rangle \\
&= \sum_{i=1}^{5} \left(\langle x_i, x_i\rangle - \langle x_i, x_{i+j}\rangle - \overline{\langle x_i, x_{i+j}\rangle} + \langle x_{i+j}, x_{i+j}\rangle\right) \\
&= \sum_{i=1}^{5} \left(2\langle x_i, x_i\rangle - \langle x_i, x_{i+j}\rangle - \overline{\langle x_i, x_{i+j}\rangle}\right) \\
&= 2S_0 - S_j - \overline{S}_j,
\end{aligned} \tag{6.9}$$

where $\overline{c}$ denotes the complex conjugation of a complex number c.

Due to (6.9), $\phi^2 = \phi + 1$ and $2\phi = 1 + \sqrt{5}$, inequality (6.7) becomes

$$\begin{aligned}
0 &\le \phi^2 \sum_{i=1}^{5} \|x_i - x_{i+1}\|^2 - \sum_{i=1}^{5} \|x_i - x_{i+2}\|^2 \\
&= \phi^2(2S_0 - S_1 - \overline{S}_1) - (2S_0 - S_2 - \overline{S}_2) \\
&= (2\phi^2 - 2)S_0 - \phi^2(S_1 + \overline{S}_1) + (S_2 + \overline{S}_2) \\
&= 2\phi S_0 - (\phi + 1)(S_1 + \overline{S}_1) + (S_2 + \overline{S}_2),
\end{aligned}$$

i.e., it is to prove that

$$\left(1 + \sqrt{5}\right)S_0 - \frac{3 + \sqrt{5}}{2}(S_1 + \overline{S}_1) + (S_2 + \overline{S}_2) \ge 0. \tag{6.10}$$

Since

$$\begin{aligned}
&\sum_{i=1}^{5} \langle x_i - x_{i+3},\, x_{i+1} - x_{i+2}\rangle \\
&= \sum_{i=1}^{5} \left(\langle x_i, x_{i+1}\rangle + \overline{\langle x_{i+2}, x_{i+3}\rangle} - \langle x_i, x_{i+2}\rangle - \overline{\langle x_{i+1}, x_{i+3}\rangle}\right) \\
&= \sum_{i=1}^{5} \langle x_i, x_{i+1}\rangle + \sum_{i=1}^{5} \overline{\langle x_i, x_{i+1}\rangle} - \sum_{i=1}^{5} \langle x_i, x_{i+2}\rangle - \sum_{i=1}^{5} \overline{\langle x_i, x_{i+2}\rangle} \\
&= (S_1 + \overline{S}_1) - (S_2 + \overline{S}_2),
\end{aligned}$$

it follows from (6.9) that

$$\begin{aligned}
0 &\le \sum_{i=1}^{5} \|x_i - x_{i+3} - \phi(x_{i+1} - x_{i+2})\|^2 \\
&= \sum_{i=1}^{5} \big\langle (x_i - x_{i+3}) - \phi(x_{i+1} - x_{i+2}),\ (x_i - x_{i+3}) - \phi(x_{i+1} - x_{i+2}) \big\rangle \\
&= \sum_{i=1}^{5} \|x_{i+3} - x_i\|^2 - \phi \sum_{i=1}^{5} \langle x_i - x_{i+3},\ x_{i+1} - x_{i+2} \rangle \\
&\quad - \phi \sum_{i=1}^{5} \overline{\langle x_i - x_{i+3},\ x_{i+1} - x_{i+2} \rangle} + \phi^2 \sum_{i=1}^{5} \|x_{i+1} - x_{i+2}\|^2 \qquad (6.11) \\
&= \sum_{i=1}^{5} \|x_i - x_{i+2}\|^2 - 2\phi(S_1 + \overline{S}_1 - S_2 - \overline{S}_2) + \phi^2 \sum_{i=1}^{5} \|x_i - x_{i+1}\|^2 \\
&= (2\phi^2 + 2)S_0 - (\phi^2 + 2\phi)S_1 - (\phi^2 + 2\phi)\overline{S}_1 \\
&\quad + (2\phi - 1)S_2 + (2\phi - 1)\overline{S}_2 \\
&= \big(5 + \sqrt{5}\big)S_0 - \frac{5 + 3\sqrt{5}}{2} S_1 - \frac{5 + 3\sqrt{5}}{2} \overline{S}_1 + \sqrt{5} S_2 + \sqrt{5} \overline{S}_2.
\end{aligned}$$

If we divide the inequality (6.11) by $\sqrt{5}$, then the resulting inequality becomes exactly the inequality (6.10), which is equivalent to our main inequality (6.7).

(*ii*) *Equality condition.* The right-hand side of (6.11) is just 0 if and only if $x_i - x_{i+3} - \phi(x_{i+1} - x_{i+2}) = 0$ for all $i \in \{1, 2, \ldots, 5\}$, which is equivalent to

$$x_{i+3} = x_i - \phi x_{i+1} + \phi x_{i+2} \qquad (6.12)$$

for all $i \in \{1, 2, \ldots, 5\}$. Assume that the right hand side of (6.11) is 0 and x_1, x_2, x_3 are given points in H. Then, by (6.12), we have

$$x_4 = x_1 - \phi x_2 + \phi x_3$$

and

$$\begin{aligned}
x_5 &= x_2 - \phi x_3 + \phi x_4 \\
&= x_2 - \phi x_3 + \phi(x_1 - \phi x_2 + \phi x_3) \\
&= \phi x_1 + (-\phi^2 + 1)x_2 + (\phi^2 - \phi)x_3 \\
&= \phi x_1 - \phi x_2 + x_3.
\end{aligned}$$

Thus, the conditions in (6.8) are true.

On the other hand, we should check under the assumptions in (6.8) that the equation (6.12) also holds when $i \in \{1, 2, \ldots, 5\}$. Indeed, our assumptions in (6.8) imply that the Eq. (6.12) holds for $i \in \{1, 2\}$ as we see in the last paragraph. It follows from (6.8) that

$$\begin{aligned} x_3 - \phi x_4 + \phi x_5 &= x_3 - \phi(x_1 - \phi x_2 + \phi x_3) + \phi(\phi x_1 - \phi x_2 + x_3) \\ &= (\phi^2 - \phi)x_1 + (\phi^2 - \phi^2)x_2 + (-\phi^2 + \phi + 1)x_3 \\ &= x_1 = x_6, \end{aligned}$$

which is just the case of (6.12) for $i = 3$, where we notice that $x_6 = x_1$, $x_7 = x_2$, and $x_8 = x_3$. Moreover, by (6.8), we have

$$\begin{aligned} x_4 - \phi x_5 + \phi x_6 &= x_4 - \phi x_5 + \phi x_1 \\ &= (x_1 - \phi x_2 + \phi x_3) - \phi(\phi x_1 - \phi x_2 + x_3) + \phi x_1 \\ &= (-\phi^2 + \phi + 1)x_1 + (\phi^2 - \phi)x_2 + (\phi - \phi)x_3 \\ &= x_2 = x_7, \end{aligned}$$

which is just the case of (6.12) for $i = 4$. Similarly, we obtain

$$\begin{aligned} x_5 - \phi x_6 + \phi x_7 &= x_5 - \phi x_1 + \phi x_2 \\ &= (\phi x_1 - \phi x_2 + x_3) - \phi x_1 + \phi x_2 \\ &= x_3 = x_8, \end{aligned}$$

which is the case of (6.12) for $i = 5$.

Hence, Eq. (6.12) holds for every $i \in \{1, 2, \ldots, 5\}$. Since x_1, x_2, and x_3 can be any points of H, the equality sign in (6.7) holds if and only if the conditions in (6.8) hold for any $x_1, x_2, x_3 \in H$. □

Remark 6.7 Let H be a real (or complex) inner product space.

(i) The number of possible distances between two of five different points of H is $\binom{5}{2} = 10$, and the number of distances contained in the inequality (6.7) is also 10.

(ii) Theorem 6.6 (ii) provides a necessary and sufficient condition to ensure the equality sign in the inequality (6.7).

(iii) In addition to the inequality (6.7), it is currently unknown whether another inequality exists that satisfies the two conditions (i) and (ii) aforementioned.

6.4 An Inequality for Distances Among n Points

From now on, let n be an integer greater than 3 and let c_n be defined as

$$c_n = \frac{\sin \frac{3\pi}{n}}{\sin \frac{\pi}{n}},$$

where the value of c_n depends on n only. We note that it follows from (6.13) below that $1 \le c_n < 3$.

Lemma 6.8 *Given an integer $n > 3$, let*

$$c_n = \frac{\sin \frac{3\pi}{n}}{\sin \frac{\pi}{n}} \quad \textit{and} \quad A_n = \begin{pmatrix} 0 & 1 & 0 \\ 0 & 0 & 1 \\ 1 & -c_n & c_n \end{pmatrix}.$$

Then $A_n^n = I$, where I denotes the 3×3 identity matrix.

Proof Since $\sin(\alpha + \beta) - \sin(\alpha - \beta) = 2\cos\alpha \sin\beta$, it holds that

$$c_n - 1 = \frac{\sin \frac{3\pi}{n}}{\sin \frac{\pi}{n}} - 1 = \frac{\sin \frac{3\pi}{n} - \sin \frac{\pi}{n}}{\sin \frac{\pi}{n}} = \frac{2\cos \frac{2\pi}{n} \sin \frac{\pi}{n}}{\sin \frac{\pi}{n}} = 2\cos \frac{2\pi}{n}. \tag{6.13}$$

Using (6.13), the characteristic polynomial of A_n, denoted by $\phi_n(t)$, is

$$\begin{aligned}
\phi_n(t) &= \det(A_n - tI) \\
&= \det \begin{pmatrix} -t & 1 & 0 \\ 0 & -t & 1 \\ 1 & -c_n & c_n - t \end{pmatrix} \\
&= -(t-1)\big(t^2 + (1 - c_n)t + 1\big) \\
&= -(t-1)\left(t^2 - 2t\cos\frac{2\pi}{n} + 1\right).
\end{aligned}$$

Let $\omega = \cos\frac{2\pi}{n} + i\sin\frac{2\pi}{n}$, and let $\overline{\omega} = \cos\frac{2\pi}{n} - i\sin\frac{2\pi}{n}$. Then the eigenvalues of A_n are 1, ω, and $\overline{\omega}$. Because the eigenvalues of A_n are all distinct, their corresponding eigenvectors are linearly independent. Thus A_n is diagonalizable.

We define P by

$$P = \begin{pmatrix} 1 & 0 & 0 \\ 0 & \omega & 0 \\ 0 & 0 & \overline{\omega} \end{pmatrix}.$$

Then there exists a matrix U satisfying $U^{-1} A_n U = P$. Hence, we obtain

$$U^{-1} A_n^n U = P^n = \begin{pmatrix} 1 & 0 & 0 \\ 0 & \omega^n & 0 \\ 0 & 0 & \overline{\omega}^n \end{pmatrix} = I.$$

Therefore, $A_n^n = U I U^{-1} = I$. □

The number of distances between any two points among the n different points is $\binom{n}{2}$, while the number of distances involved in the inequality (6.14) is $3n$, which is not equal to $\binom{n}{2}$ unless $n = 7$. Nevertheless, the inequality (6.14) is interesting and important enough. Indeed, the following theorem was proved by S.-M. Jung and D. Nam [38].

Theorem 6.9 (Jung and Nam) *Given an integer $n > 3$, let $c_n = \frac{\sin \frac{3\pi}{n}}{\sin \frac{\pi}{n}}$ and H a real (or complex) inner product space.*

(*i*) *If n points $x_1, x_2, \ldots, x_n$ are arbitrarily given in H, then the following inequality holds:*

$$\begin{aligned} \left(c_n^2 + 2c_n\right) \sum_{i=1}^{n} \|x_i - x_{i+1}\|^2 + \sum_{i=1}^{n} \|x_i - x_{i+3}\|^2 \\ \geq 2c_n \sum_{i=1}^{n} \|x_i - x_{i+2}\|^2, \end{aligned} \tag{6.14}$$

where $x_{n+1} = x_1$, $x_{n+2} = x_2$, and $x_{n+3} = x_3$ for notational convenience.

(*ii*) *The equality sign in* (6.14) *holds if and only if for previously given points $x_1, x_2, x_3 \in H$, the points $x_4, x_5, \ldots, x_n$ are determined by the recursion formula*

$$x_{i+3} = x_i - c_n x_{i+1} + c_n x_{i+2} \tag{6.15}$$

for all $i \in \{1, 2, \ldots, n-3\}$.

Proof Because the proof of this theorem for real inner product spaces is similar as the complex case, we will prove this theorem when H is a complex inner product space.

(i) We set

$$S_j = \sum_{i=1}^{n} \langle x_i, x_{i+j} \rangle$$

for each $j \in \{0, 1, 2, 3\}$. Then, as we did in (6.9), we obtain

$$\begin{aligned}
&\sum_{i=1}^{n} \|x_i - x_{i+j}\|^2 \\
&\quad = \sum_{i=1}^{n} \langle x_i - x_{i+j},\ x_i - x_{i+j} \rangle \\
&\quad = \sum_{i=1}^{n} \left(\langle x_i, x_i \rangle - \langle x_i, x_{i+j} \rangle - \overline{\langle x_i, x_{i+j} \rangle} + \langle x_{i+j}, x_{i+j} \rangle \right) \\
&\quad = \sum_{i=1}^{n} \left(2\langle x_i, x_i \rangle - \langle x_i, x_{i+j} \rangle - \overline{\langle x_i, x_{i+j} \rangle} \right) \\
&\quad = 2S_0 - S_j - \overline{S}_j,
\end{aligned} \tag{6.16}$$

for each $j \in \{0, 1, 2, 3\}$, where $\overline{c}$ denotes the complex conjugation of a complex number c.

Since

$$\begin{aligned}
&\sum_{i=1}^{n} \langle x_i - x_{i+3},\ x_{i+1} - x_{i+2} \rangle \\
&\quad = \sum_{i=1}^{n} \left(\langle x_i, x_{i+1} \rangle + \overline{\langle x_{i+2}, x_{i+3} \rangle} - \langle x_i, x_{i+2} \rangle - \overline{\langle x_{i+1}, x_{i+3} \rangle} \right) \\
&\quad = \sum_{i=1}^{n} \langle x_i, x_{i+1} \rangle + \sum_{i=1}^{n} \overline{\langle x_i, x_{i+1} \rangle} - \sum_{i=1}^{n} \langle x_i, x_{i+2} \rangle - \sum_{i=1}^{n} \overline{\langle x_i, x_{i+2} \rangle} \\
&\quad = (S_1 + \overline{S}_1) - (S_2 + \overline{S}_2),
\end{aligned} \tag{6.17}$$

it follows that

$$\sum_{i=1}^{n} \overline{\langle x_i - x_{i+3},\ x_{i+1} - x_{i+2} \rangle} = (S_1 + \overline{S}_1) - (S_2 + \overline{S}_2), \tag{6.18}$$

and by (6.16), (6.17) and (6.18), the following inequality holds:

$$
\begin{aligned}
0 &\leq \sum_{i=1}^{n} \|x_i - x_{i+3} - c_n(x_{i+1} - x_{i+2})\|^2 \\
&= \sum_{i=1}^{n} \big\langle (x_i - x_{i+3}) - c_n(x_{i+1} - x_{i+2}),\ (x_i - x_{i+3}) - c_n(x_{i+1} - x_{i+2}) \big\rangle \\
&= \sum_{i=1}^{n} \|x_i - x_{i+3}\|^2 - c_n \sum_{i=1}^{n} \langle x_i - x_{i+3},\ x_{i+1} - x_{i+2} \rangle \\
&\quad - c_n \sum_{i=1}^{n} \overline{\langle x_i - x_{i+3},\ x_{i+1} - x_{i+2} \rangle} + c_n^2 \sum_{i=1}^{n} \|x_{i+1} - x_{i+2}\|^2 \\
&= \sum_{i=1}^{n} \|x_i - x_{i+3}\|^2 - 2c_n(S_1 + \overline{S}_1 - S_2 - \overline{S}_2) + c_n^2 \sum_{i=1}^{n} \|x_i - x_{i+1}\|^2 \\
&= \sum_{i=1}^{n} \|x_i - x_{i+3}\|^2 - 2c_n\big((2S_0 - S_2 - \overline{S}_2) - (2S_0 - S_1 - \overline{S}_1)\big) \qquad (6.19) \\
&\quad + c_n^2 \sum_{i=1}^{n} \|x_i - x_{i+1}\|^2 \\
&= \sum_{i=1}^{n} \|x_i - x_{i+3}\|^2 - 2c_n \sum_{i=1}^{n} \|x_i - x_{i+2}\|^2 + 2c_n \sum_{i=1}^{n} \|x_i - x_{i+1}\|^2 \\
&\quad + c_n^2 \sum_{i=1}^{n} \|x_i - x_{i+1}\|^2 \\
&= (c_n^2 + 2c_n) \sum_{i=1}^{n} \|x_i - x_{i+1}\|^2 + \sum_{i=1}^{n} \|x_i - x_{i+3}\|^2 \\
&\quad - 2c_n \sum_{i=1}^{n} \|x_i - x_{i+2}\|^2.
\end{aligned}
$$

Hence, inequality (6.14) is true.

(ii) *Equality condition*. We choose arbitrary $x_1, x_2, x_3 \in H$ and determine $x_4, x_5, \ldots, x_n$ recursively by substituting i with $1, 2, \ldots, n-3$ in (6.15). Then, with the same A_n in Lemma 6.8, it follows from (6.15) that

$$
A_n \begin{pmatrix} x_i \\ x_{i+1} \\ x_{i+2} \end{pmatrix} = \begin{pmatrix} 0 & 1 & 0 \\ 0 & 0 & 1 \\ 1 & -c_n & c_n \end{pmatrix} \begin{pmatrix} x_i \\ x_{i+1} \\ x_{i+2} \end{pmatrix} = \begin{pmatrix} x_{i+1} \\ x_{i+2} \\ x_i - c_n x_{i+1} + c_n x_{i+2} \end{pmatrix} = \begin{pmatrix} x_{i+1} \\ x_{i+2} \\ x_{i+3} \end{pmatrix}
$$

for all $i \in \{1, 2, \ldots, n-3\}$.

Temporarily, we set

$$y_1 = x_{n-2} - c_n x_{n-1} + c_n x_n$$

$$y_2 = x_{n-1} - c_n x_n + c_n y_1$$

$$y_3 = x_n - c_n y_1 + c_n y_2.$$

Since $A_n^n = I$ by Lemma 6.8, it holds that

$$\begin{aligned}
\begin{pmatrix} x_1 \\ x_2 \\ x_3 \end{pmatrix} &= A_n^n \begin{pmatrix} x_1 \\ x_2 \\ x_3 \end{pmatrix} = A_n^3 A_n^{n-3} \begin{pmatrix} x_1 \\ x_2 \\ x_3 \end{pmatrix} \\
&= A_n^3 \begin{pmatrix} x_{n-2} \\ x_{n-1} \\ x_n \end{pmatrix} = A_n^2 \begin{pmatrix} x_{n-1} \\ x_n \\ x_{n-2} - c_n x_{n-1} + c_n x_n \end{pmatrix} \\
&= A_n^2 \begin{pmatrix} x_{n-1} \\ x_n \\ y_1 \end{pmatrix} = A_n \begin{pmatrix} x_n \\ y_1 \\ x_{n-1} - c_n x_n + c_n y_1 \end{pmatrix} \\
&= A_n \begin{pmatrix} x_n \\ y_1 \\ y_2 \end{pmatrix} = \begin{pmatrix} y_1 \\ y_2 \\ x_n - c_n y_1 + c_n y_2 \end{pmatrix} = \begin{pmatrix} y_1 \\ y_2 \\ y_3 \end{pmatrix}.
\end{aligned}$$

Therefore, $y_1 = x_1$, $y_2 = x_2$, $y_3 = x_3$, and

$$x_1 = x_{n-2} - c_n x_{n-1} + c_n x_n$$

$$x_2 = x_{n-1} - c_n x_n + c_n x_1$$

$$x_3 = x_n - c_n x_1 + c_n x_2.$$

Hence, condition (6.15) is satisfied for each $i \in \{1, 2, \ldots, n\}$.

Finally, it is obvious that the right hand side of (6.19) is zero if and only if condition (6.15) is satisfied for all $i \in \{1, 2, \ldots, n\}$. □

When studying inequalities for distances between any two points among given n points, it is advisable to adhere to the following two recommendations:

(†) All possible distances between those n points are included in the inequality.
(‡) The necessary and sufficient condition for the equality sign is presented.

Remark 6.10 Let H be a real (or complex) inner product space and $n > 3$.

(i) The number of distances between any two points among the n different points is $\binom{n}{2}$, while the number of distances involved in the inequality (6.14) is $3n$, which is not equal to $\binom{n}{2}$ unless $n = 7$.

(ii) Theorem 6.9 (ii) provides a necessary and sufficient condition to ensure the equality sign in the inequality (6.14).

(iii) In this section, for each $n \in \{4, 5, 6, 7, 8, 10, 12, \ldots\}$, we have presented an inequality that satisfies both conditions (†) and (‡).

(iv) If $n \in \{9, 11, 13, \ldots\}$, it is not yet known whether an inequality exists that satisfies both conditions (†) and (‡).

Jung, Lee, and Nam's Partial Solutions

7

Abstract

In this chapter, we discuss ideas for partially solving the Aleksandrov-Rassias problems using the inequalities presented in Chap. 6. In the first section, we partially solve the Aleksandrov-Rassias problems by using the inequality for the distances among six points presented in Sect. 6.1. Section 7.2 is devoted to proving that any mapping between real Hilbert spaces whose dimensions are greater than 2 is an affine isometry if the distance 1 is preserved, $\frac{1}{\sqrt{2}}$ is contractive, and when $\sqrt{3}$ is extensive. In Sect. 7.3, we give a partial solution to the Aleksandrov-Rassias problems by proving that when the distance 1 is contractive and the golden ratio is extensive by a mapping defined between real Hilbert spaces and when the dimension of its domain is greater than 2, then this mapping is an affine isometry. In the last section of this chapter, we prove that a mapping between real Hilbert spaces whose domain has the dimension greater than 2 is an affine isometry if the distances 1 and α are contractive, β is extensive, and if the distances 1, α and β satisfy some suitable conditions. The main results presented in this chapter have been extracted from the papers by Jung (Nonlinear Anal 62(4):675–681, 2005); Jung and Lee (J Math Anal Appl 324(2):1363–1369, 2006); Jung and Nam (J Math Inequal 12(4):1189–1199, 2018); Jung and Nam (J Math Inequal 13(4):969–981, 2019) and explained in detail so that the reader can easily understand them.

7.1 Applications of an Inequality for Six Points

A distance $\rho > 0$ is said to be contractive (or non-expanding) by a mapping $f : X \to Y$ between normed spaces if and only if $\|f(x) - f(y)\| \leq \rho$ for all $x, y \in X$ with $\|x - y\| = \rho$. Similarly, a distance ρ is said to be extensive (or nonshrinking) by f if and only if the inequality $\|f(x) - f(y)\| \geq \rho$ holds for all $x, y \in X$ with $\|x - y\| = \rho$. We say that ρ is

S.-M. Jung, *Aleksandrov-Rassias Problems on Distance Preserving Mappings*, Frontiers in Mathematics, https://doi.org/10.1007/978-3-031-77613-7_7

preserved (or conservative) by f if and only if ρ is both contractive and extensive by f. If f is an isometry, every distance $\rho > 0$ is preserved by f, and conversely.

First, using the short diagonals lemma and the parallelogram law, we study the Aleksandrov-Rassias problems when two distances are contractive and another one is extensive by a mapping. The following theorem was proved by S.-M. Jung [27].

Theorem 7.1 (Jung) *Let $\mathbb{E}^n$ be an n-dimensional Euclidean space, where n is an integer greater than 1. If the distances $\rho, \sigma > 0$ are contractive and the distance $\sqrt{\rho^2 + \sigma^2}$ is extensive by a mapping $f : \mathbb{E}^n \to \mathbb{E}^n$, then f is an affine isometry.*

Proof As before, we denote by $\|u - v\|$ the distance between two points $u, v \in \mathbb{E}^n$, *i.e.*, $\|u - v\|^2 = \sum_{i=1}^{n} (u_i - v_i)^2$, where $u = (u_1, u_2, \ldots, u_n)$ and $v = (v_1, v_2, \ldots, v_n)$. We consider a rectangle (a quadrilateral with four right angles) with $\|o - p\| = \|q - r\| = \rho$ and $\|p - q\| = \|r - o\| = \sigma$.

It follows from the parallelogram law that $\|o - q\| = \|p - r\| = \sqrt{\rho^2 + \sigma^2}$. Since the distance $\sqrt{\rho^2 + \sigma^2}$ is extensive, we have

$$\|f(o) - f(q)\|^2 + \|f(p) - f(r)\|^2 \geq \|o - q\|^2 + \|p - r\|^2. \tag{7.1}$$

Due to the fact that the points o, p, q, r comprise the vertices of a rectangle such that $o - q$ and $p - r$ are the diagonals, the parallelogram law implies that

$$\|o - q\|^2 + \|p - r\|^2 = \|o - p\|^2 + \|p - q\|^2 + \|q - r\|^2 + \|r - o\|^2. \tag{7.2}$$

Moreover, since both ρ and σ are assumed to be contractive, it holds that

$$\begin{aligned} &\|o - p\|^2 + \|p - q\|^2 + \|q - r\|^2 + \|r - o\|^2 \\ &\geq \|f(o) - f(p)\|^2 + \|f(p) - f(q)\|^2 \\ &\quad + \|f(q) - f(r)\|^2 + \|f(r) - f(o)\|^2. \end{aligned} \tag{7.3}$$

By the short diagonals lemma, we obtain

$$\begin{aligned} &\|f(o) - f(p)\|^2 + \|f(p) - f(q)\|^2 \\ &+ \|f(q) - f(r)\|^2 + \|f(r) - f(o)\|^2 \\ &\geq \|f(o) - f(q)\|^2 + \|f(p) - f(r)\|^2. \end{aligned} \tag{7.4}$$

Altogether, combining (7.1), (7.2), (7.3), and (7.4), we have

$$\|o - q\|^2 + \|p - r\|^2 = \|f(o) - f(q)\|^2 + \|f(p) - f(r)\|^2 \tag{7.5}$$

and

$$\begin{aligned}
&\|o - p\|^2 + \|p - q\|^2 + \|q - r\|^2 + \|r - o\|^2 \\
&= \|f(o) - f(p)\|^2 + \|f(p) - f(q)\|^2 \\
&\quad + \|f(q) - f(r)\|^2 + \|f(r) - f(o)\|^2.
\end{aligned} \tag{7.6}$$

Our hypotheses obviously imply that

$$\begin{aligned}
&\|o - q\| \le \|f(o) - f(q)\|, \ \|p - r\| \le \|f(p) - f(r)\|, \\
&\|o - p\| \ge \|f(o) - f(p)\|, \ \|p - q\| \ge \|f(p) - f(q)\|, \\
&\|q - r\| \ge \|f(q) - f(r)\|, \ \|r - o\| \ge \|f(r) - f(o)\|.
\end{aligned} \tag{7.7}$$

It follows from (7.5), (7.6), and (7.7) that

$$\begin{aligned}
&\|f(o) - f(q)\| = \|f(p) - f(r)\| = \sqrt{\rho^2 + \sigma^2}, \\
&\|f(o) - f(p)\| = \|f(q) - f(r)\| = \rho, \\
&\|f(p) - f(q)\| = \|f(r) - f(o)\| = \sigma.
\end{aligned}$$

For example, we can check $\|f(o) - f(q)\| = \|o - q\| = \sqrt{\rho^2 + \sigma^2}$ using (7.5) and (7.7): If $\|f(o) - f(q)\| > \|o - q\|$, it would then follow from (7.7) that

$$\|f(o) - f(q)\|^2 + \|f(p) - f(r)\|^2 > \|o - q\|^2 + \|p - r\|^2,$$

which would contradict (7.5). Hence, considering (7.7), we can conclude that $\|f(o) - f(q)\| = \|o - q\| = \sqrt{\rho^2 + \sigma^2}$.

For any given points $o, p \in \mathbb{E}^n$ with $\|o - p\| = \rho$, we can select two points $q, r \in \mathbb{E}^n$ such that the four points o, p, q, r comprise the vertices of a rectangle, as shown in Fig. 7.1. On account of the above argument, we may conclude that $\|f(o) - f(p)\| = \rho$. Therefore, f preserves the distance ρ. (To prove that f preserves the other distances σ and $\sqrt{\rho^2 + \sigma^2}$, we could make a similar argument to the one above, but this is not necessary. Because we will apply the Beckman-Quarles theorem, for which it is enough to prove that f preserves the distance ρ.) Finally, we apply the Beckman-Quarles theorem to our case and conclude that f is an affine isometry. □

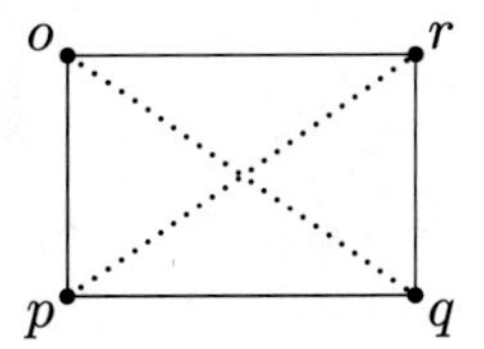

Fig. 7.1 The four points o, p, q, r are the vertices of a rectangle with $\|o - p\| = \|q - r\| = \rho$ and $\|p - q\| = \|r - o\| = \sigma$

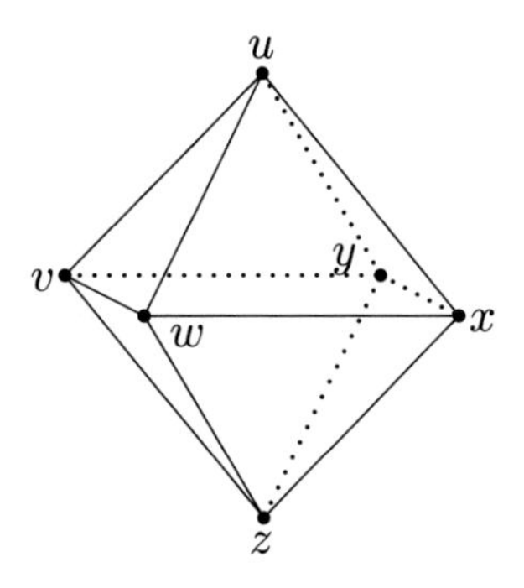

Fig. 7.2 The six points u, v, w, x, y, z are the vertices of the octahedron. Each four points $\{u, v, z, x\}$, $\{u, w, z, y\}$, and $\{v, w, x, y\}$ is the vertices of the corresponding parallelogram, respectively

We assume that H_1 is a real (or complex) inner product space with $\dim H_1 > 2$ and there exist six points $u, v, w, x, y, z \in H_1$ with the property that each pair of four points $\{u, v, z, x\}$, $\{u, w, z, y\}$, and $\{v, w, x, y\}$ determines a corresponding parallelogram, respectively, as we see in Fig. 7.2:
We may say that the points u, v, w, x, y, z determine the octahedron in Fig. 7.2.

In addition, all distances among the six points u, v, w, x, y, z are assumed to be

$$\begin{aligned}
&\|v-u\| = \|x-z\| = a_1, \quad \|z-v\| = \|u-x\| = a_2, \\
&\|w-u\| = \|y-z\| = b_1, \quad \|z-w\| = \|u-y\| = b_2, \\
&\|w-v\| = \|y-x\| = c_1, \quad \|x-w\| = \|v-y\| = c_2, \\
&\|z-u\| = \alpha, \quad \|y-w\| = \beta, \quad \|x-v\| = \gamma,
\end{aligned} \tag{7.8}$$

where $a_1, a_2, b_1, b_2, c_1, c_2, \alpha, \beta, \gamma$ are some positive real numbers.

Theorem 7.2 (Jung) *Let H_1 and H_2 be real (or complex) inner product spaces with $\dim H_1 > 2$ and $\dim H_2 > 2$. Assume that the distances a_1, a_2, b_1, b_2, c_1, and c_2 are contractive and the distances α, β, and γ are extensive by a mapping $f : H_1 \to H_2$, where a_1, a_2, b_1, b_2, c_1, c_2, α, β, and γ are assumed to satisfy all the conditions in (7.8) for the octahedron in Fig. 7.2. Then, f preserves the distances a_1, a_2, b_1, b_2, c_1, c_2, α, β, and γ.*

Proof Assume that the points $u, v, w, x, y, z \in H_1$ determine the octahedron shown in Fig. 7.2 and satisfy the conditions of (7.8).

Since $\|z-u\| = \alpha$, $\|y-w\| = \beta$, $\|x-v\| = \gamma$ by (7.8), and α, β, and γ are extensive by f, we have

$$\begin{aligned}
&2\big(\|f(z)-f(u)\|^2 + \|f(x)-f(v)\|^2 + \|f(y)-f(w)\|^2\big) \\
&\quad \geq 2\big(\|z-u\|^2 + \|x-v\|^2 + \|y-w\|^2\big).
\end{aligned} \tag{7.9}$$

Since each pair of four points $\{u, v, z, x\}$, $\{u, w, z, y\}$, and $\{v, w, x, y\}$ comprises the vertices of the corresponding parallelogram presented in Fig. 7.2, it follows from Theorem 6.2 (ii) that

$$
\begin{aligned}
&2\big(\|z-u\|^2+\|x-v\|^2+\|y-w\|^2\big)\\
&\quad=\|v-u\|^2+\|z-v\|^2+\|x-z\|^2+\|u-x\|^2\\
&\qquad+\|w-v\|^2+\|x-w\|^2+\|y-x\|^2+\|v-y\|^2\\
&\qquad+\|w-u\|^2+\|z-w\|^2+\|y-z\|^2+\|u-y\|^2.
\end{aligned}
\tag{7.10}
$$

Moreover, since the distances a_1, a_2, b_1, b_2, c_1, and c_2 are contractive, it follows from (7.8) and Theorem 6.2 (i) that

$$
\begin{aligned}
&\|v-u\|^2+\|z-v\|^2+\|x-z\|^2+\|u-x\|^2\\
&+\|w-v\|^2+\|x-w\|^2+\|y-x\|^2+\|v-y\|^2\\
&+\|w-u\|^2+\|z-w\|^2+\|y-z\|^2+\|u-y\|^2\\
&\quad\ge\|f(v)-f(u)\|^2+\|f(z)-f(v)\|^2+\|f(x)-f(z)\|^2\\
&\qquad+\|f(u)-f(x)\|^2+\|f(w)-f(v)\|^2+\|f(x)-f(w)\|^2\\
&\qquad+\|f(y)-f(x)\|^2+\|f(v)-f(y)\|^2+\|f(w)-f(u)\|^2\\
&\qquad+\|f(z)-f(w)\|^2+\|f(y)-f(z)\|^2+\|f(u)-f(y)\|^2\\
&\quad\ge 2\big(\|f(z)-f(u)\|^2+\|f(x)-f(v)\|^2+\|f(y)-f(w)\|^2\big).
\end{aligned}
\tag{7.11}
$$

Since our hypotheses imply

$$
\begin{aligned}
&\|z-u\|\le\|f(z)-f(u)\|, && \|x-v\|\le\|f(x)-f(v)\|,\\
&\|y-w\|\le\|f(y)-f(w)\|, && \|v-u\|\ge\|f(v)-f(u)\|,\\
&\|z-v\|\ge\|f(z)-f(v)\|, && \|x-z\|\ge\|f(x)-f(z)\|,\\
&\|u-x\|\ge\|f(u)-f(x)\|, && \|w-v\|\ge\|f(w)-f(v)\|,\\
&\|x-w\|\ge\|f(x)-f(w)\|, && \|y-x\|\ge\|f(y)-f(x)\|,\\
&\|v-y\|\ge\|f(v)-f(y)\|, && \|w-u\|\ge\|f(w)-f(u)\|,\\
&\|z-w\|\ge\|f(z)-f(w)\|, && \|y-z\|\ge\|f(y)-f(z)\|,\\
&\|u-y\|\ge\|f(u)-f(y)\|,
\end{aligned}
\tag{7.12}
$$

it follows from (7.9), (7.10), (7.11), and (7.12) that

$$
\begin{aligned}
&\|f(v)-f(u)\|=\|f(x)-f(z)\|=a_1, && \|f(z)-f(u)\|=\alpha,\\
&\|f(z)-f(v)\|=\|f(u)-f(x)\|=a_2, && \|f(y)-f(w)\|=\beta,\\
&\|f(w)-f(u)\|=\|f(y)-f(z)\|=b_1, && \|f(x)-f(v)\|=\gamma\\
&\|f(z)-f(w)\|=\|f(u)-f(y)\|=b_2,\\
&\|f(w)-f(v)\|=\|f(y)-f(x)\|=c_1,\\
&\|f(x)-f(w)\|=\|f(v)-f(y)\|=c_2.
\end{aligned}
$$

For any given $u, v \in H_1$ with $\|v - u\| = a_1$, we can select four points w, x, y, z in H_1 such that u, v, w, x, y, z determine the octahedron, as shown in Fig. 7.2. Due to the above argument, we can conclude that $\|f(v) - f(u)\| = a_1$. For other distances such as a_2, b_1, b_2, c_1, c_2, α, β, and γ, we can apply a similar argument. Therefore, f preserves all the distances $a_1, a_2, b_1, b_2, c_1, c_2, \alpha, \beta$, and γ. □

If $a_1 = a_2 = b_1 = b_2 = c_1 = c_2 = 1$ and $\alpha = \beta = \gamma = \sqrt{2}$, the six points u, v, w, x, y, z determine the unit regular octahedron in Fig. 7.2. For this case, it follows from Theorem 7.2 that f preserves the distances 1 and $\sqrt{2}$. Due to Theorem 5.3, the following statement is true.

Corollary 7.3 *Let H_1 and H_2 be real Hilbert spaces with* $\dim H_1 > 2$ *and* $\dim H_2 > 2$. *Assume that the distance* 1 *is contractive and the distance* $\sqrt{2}$ *is extensive by a mapping* $f : H_1 \to H_2$. *Then, f is an affine isometry.*

Remark 7.4 In Theorem 7.2 and Corollary 7.3, H_1 and H_2 are allowed to be infinite-dimensional inner product spaces (or Hilbert spaces), while we assume in Theorem 7.1 that $\mathbb{E}^n$ is a finite-dimensional Euclidean space.

7.2 Applications of an Inequality for $2n$ Points

We label the vertices of an arbitrary (possibly degenerate) parallelogram with p_{11}, p_{12}, p_{21}, and p_{22}, as we see on the left side of Fig. 7.3 . Similarly, we label the vertices of any (possibly degenerate) octahedron with p_{11}, p_{12}, p_{21}, p_{22}, p_{31}, and p_{32} as shown on the right-hand side of Fig. 7.3.

We continue this construction by extending it to the general case. Suppose we have constructed an n-dimensional polyhedron with $2n$ vertices, $p_{11}, p_{12}, \ldots, p_{n1}, p_{n2}$. Now, we add two more points, denoted by $p_{(n+1)1}$ and $p_{(n+1)2}$, to construct an $(n + 1)$-dimensional polyhedron in the following manner: Each of the new points, $p_{(n+1)1}$ and $p_{(n+1)2}$, is connected to the existing $2n$ vertices, $p_{11}, p_{12}, \ldots, p_{n1}, p_{n2}$.

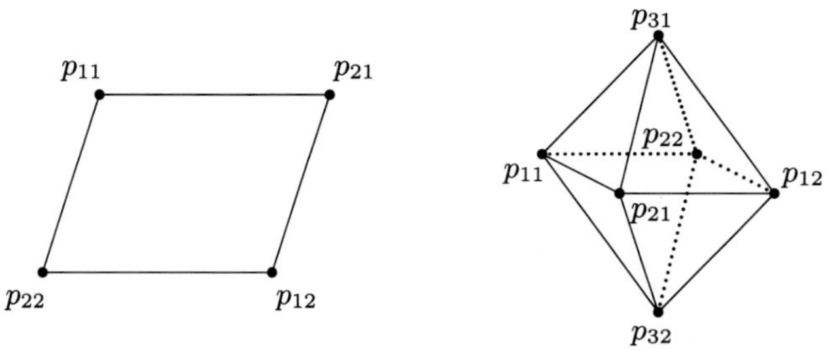

Fig. 7.3 The vertices of any n-dimensional polyhedron are denoted as $p_{11}, p_{12}, \ldots, p_{n1}, p_{n2}$, where p_{i1} and p_{i2} are opposite vertices for all $i \in \{1, 2, \ldots, n\}$

For the n-dimensional polyhedron constructed as above, we will denote its $2n$ vertices by $p_{11}, p_{12}, \ldots, p_{n1}, p_{n2}$ as the above construction. We define

$$\alpha_{ij} = \|p_{i1} - p_{j1}\|, \quad \beta_{ij} = \|p_{i2} - p_{j2}\|, \quad \gamma_{ij} = \|p_{i1} - p_{j2}\|$$

for all $i, j \in \{1, 2, \ldots, n\}$.

In the following theorem, which was proved in a paper [36] by S.-M. Jung and K.-S. Lee, we assume that for any $i, j \in \{1, 2, \ldots, n\}$ with $i < j$, each pair of four points, $\{p_{i1}, p_{i2}, p_{j1}, p_{j2}\}$, comprises the vertices of a corresponding parallelogram.

Theorem 7.5 (Jung and Lee) *Let H_1 and H_2 be real (or complex) inner product spaces with $\dim H_1 \geq n$ and $\dim H_2 \geq n$, where n is an integer greater than 1. Assume that the distances α_{ij}, β_{ij}, and γ_{ij} are contractive by a mapping $f : H_1 \to H_2$ for all $i, j \in \{1, 2, \ldots, n\}$ with $i < j$ and that the distances γ_{ii} are extensive by f for each $i \in \{1, 2, \ldots, n\}$. Then, f preserves all the distances α_{ij}, β_{ij}, and γ_{ii} for all $i, j \in \{1, 2, \ldots, n\}$ with $i < j$.*

Proof According to the short diagonals lemma, the inequality

$$\sum_{k,\ell\in\{1,2\}} \|p_{ik} - p_{j\ell}\|^2 \geq \|p_{i1} - p_{i2}\|^2 + \|p_{j1} - p_{j2}\|^2$$

holds for any $i, j \in \{1, 2, \ldots, n\}$ with $i < j$. Furthermore, it holds that

$$\begin{aligned}
&\sum_{\substack{i,j\in\{1,2,\ldots,n\}\\ i<j}} \left(\|p_{i1} - p_{i2}\|^2 + \|p_{j1} - p_{j2}\|^2\right) \\
&= \sum_{i=1}^{n-1}\sum_{j=i+1}^{n} \left(\|p_{i1} - p_{i2}\|^2 + \|p_{j1} - p_{j2}\|^2\right) \\
&= \sum_{i=1}^{n-1}(n-i)\|p_{i1} - p_{i2}\|^2 + \sum_{i=1}^{n-1}\sum_{j=i+1}^{n} \|p_{j1} - p_{j2}\|^2 \\
&= \sum_{i=1}^{n-1}(n-i)\|p_{i1} - p_{i2}\|^2 + \sum_{j=2}^{n}\sum_{i=1}^{j-1} \|p_{j1} - p_{j2}\|^2 \\
&= \sum_{i=1}^{n}(n-i)\|p_{i1} - p_{i2}\|^2 + \sum_{j=1}^{n}(j-1)\|p_{j1} - p_{j2}\|^2 \\
&= (n-1)\sum_{i\in\{1,2,\ldots,n\}} \|p_{i1} - p_{i2}\|^2.
\end{aligned} \tag{7.13}$$

Since the distances $\gamma_{ii} = \|p_{i1} - p_{i2}\|$ are extensive by f for all $i \in \{1, 2, \ldots, n\}$ and each pair of four points $\{p_{i1}, p_{i2}, p_{j1}, p_{j2}\}$ comprises the vertices of a corresponding parallelogram for any $i, j \in \{1, 2, \ldots, n\}$ with $i < j$, it follows from (7.13) and the parallelogram law that

$$\begin{aligned}
(n-1) & \sum_{i\in\{1,2,\ldots,n\}} \|f(p_{i1}) - f(p_{i2})\|^2 \\
&\geq (n-1) \sum_{i\in\{1,2,\ldots,n\}} \|p_{i1} - p_{i2}\|^2 \\
&= \sum_{\substack{i,j \in \{1,2,\ldots,n\} \\ i<j}} \left(\|p_{i1} - p_{i2}\|^2 + \|p_{j1} - p_{j2}\|^2\right) \\
&= \sum_{\substack{i,j \in \{1,2,\ldots,n\} \\ i<j}} \sum_{k,\ell\in\{1,2\}} \|p_{ik} - p_{j\ell}\|^2 \\
&= \sum_{\substack{i,j \in \{1,2,\ldots,n\} \\ k,\ell \in \{1,2\} \\ i<j}} \|p_{ik} - p_{j\ell}\|^2.
\end{aligned}$$

(We recall that the set of four points $\{p_{i1}, p_{i2}, p_{j1}, p_{j2}\}$ comprises the vertices of a parallelogram and that p_{i1} and p_{j1} are the opposite vertices to p_{i2} and p_{j2}, respectively.)

We note that the distances α_{ij}, β_{ij}, and γ_{ij} are contractive by f for all $i, j \in \{1, 2, \ldots, n\}$ with $i < j$ and that the distances γ_{ii} are extensive by f for any $i \in \{1, 2, \ldots, n\}$. Since p_{i1}, p_{i2}, p_{j1}, and p_{j2} comprise the vertices of a parallelogram, it follows that $\|p_{i2} - p_{j1}\| = \|p_{i1} - p_{j2}\| = \gamma_{ij}$. Hence, it follows from the last inequality that

$$\begin{aligned}
(n-1) & \sum_{i\in\{1,2,\ldots,n\}} \|f(p_{i1}) - f(p_{i2})\|^2 \\
&\geq (n-1) \sum_{i\in\{1,2,\ldots,n\}} \|p_{i1} - p_{i2}\|^2 \\
&= \sum_{\substack{i,j \in \{1,2,\ldots,n\} \\ k,\ell \in \{1,2\} \\ i<j}} \|p_{ik} - p_{j\ell}\|^2 \\
&\geq \sum_{\substack{i,j \in \{1,2,\ldots,n\} \\ k,\ell \in \{1,2\} \\ i<j}} \|f(p_{ik}) - f(p_{j\ell})\|^2 \\
&\geq (n-1) \sum_{i\in\{1,2,\ldots,n\}} \|f(p_{i1}) - f(p_{i2})\|^2,
\end{aligned}$$

where the last inequality follows from Theorem 6.4 (i).

From the first two lines of the last inequality, it follows that

$$\sum_{i\in\{1,2,\ldots,n\}} \|f(p_{i1}) - f(p_{i2})\|^2 = \sum_{i\in\{1,2,\ldots,n\}} \|p_{i1} - p_{i2}\|^2. \tag{7.14}$$

In addition, it follows from the third and fourth lines of the last inequality that

$$\sum_{\substack{i,\, j \in \{1,2,\ldots,n\} \\ k,\ell \in \{1,2\} \\ i<j}} \|p_{ik} - p_{j\ell}\|^2 = \sum_{\substack{i,\, j \in \{1,2,\ldots,n\} \\ k,\ell \in \{1,2\} \\ i<j}} \|f(p_{ik}) - f(p_{j\ell})\|^2. \tag{7.15}$$

Since $\|f(p_{i1}) - f(p_{i2})\| \geq \|p_{i1} - p_{i2}\|$ and $\|p_{ik} - p_{j\ell}\| \geq \|f(p_{ik}) - f(p_{j\ell})\|$ for all $i, j \in \{1, 2, \ldots, n\}$ with $i < j$ and $k, \ell \in \{1, 2\}$, it follows from (7.14) and (7.15) that

$$\|f(p_{i1}) - f(p_{i2})\| = \|p_{i1} - p_{i2}\| = \gamma_{ii},$$

$$\|f(p_{ik}) - f(p_{j\ell})\| = \|p_{ik} - p_{j\ell}\| = \begin{cases} \alpha_{ij} & (\text{for } k = \ell = 1), \\ \beta_{ij} & (\text{for } k = \ell = 2), \\ \gamma_{ij} & (\text{for } k = 1 \text{ and } \ell = 2), \\ \gamma_{ij} & (\text{for } k = 2 \text{ and } \ell = 1) \end{cases}$$

for all $i, j \in \{1, 2, \ldots, n\}$ with $i < j$, which completes our proof. □

We now deal with the case of $n = 3$. Let H_1 be a real Hilbert space with $\dim H_1 > 2$. We consider the octahedron in H_1 which is determined by

$$\begin{aligned} p_{11} &= \left(\tfrac{\sqrt{3}}{2}, 0, 0, 0, \ldots, 0\right), & p_{12} &= \left(-\tfrac{\sqrt{3}}{2}, 0, 0, 0, \ldots, 0\right), \\ p_{21} &= \left(0, \tfrac{1}{2}, 0, 0, \ldots, 0\right), & p_{22} &= \left(0, -\tfrac{1}{2}, 0, 0, \ldots, 0\right), \\ p_{31} &= \left(0, 0, \tfrac{1}{2}, 0, \ldots, 0\right), & p_{32} &= \left(0, 0, -\tfrac{1}{2}, 0, \ldots, 0\right). \end{aligned}$$

In this case, we have

$$\begin{aligned} \alpha_{12} &= \|p_{11} - p_{21}\| = 1, & \alpha_{13} &= \|p_{11} - p_{31}\| = 1, \\ \alpha_{23} &= \|p_{21} - p_{31}\| = \tfrac{1}{\sqrt{2}}, & \beta_{12} &= \|p_{12} - p_{22}\| = 1, \\ \beta_{13} &= \|p_{12} - p_{32}\| = 1, & \beta_{23} &= \|p_{22} - p_{32}\| = \tfrac{1}{\sqrt{2}}, \\ \gamma_{12} &= \|p_{11} - p_{22}\| = 1, & \gamma_{13} &= \|p_{11} - p_{32}\| = 1, \\ \gamma_{23} &= \|p_{21} - p_{32}\| = \tfrac{1}{\sqrt{2}}, & \gamma_{11} &= \|p_{11} - p_{12}\| = \sqrt{3}, \\ \gamma_{22} &= \|p_{21} - p_{22}\| = 1, & \gamma_{33} &= \|p_{31} - p_{32}\| = 1. \end{aligned}$$

Using Theorems 5.1 and 7.5, we can easily prove the following corollary.

Corollary 7.6 *Let H_1 and H_2 be real Hilbert spaces with $\dim H_1 > 2$ and $\dim H_2 > 2$. If the distance 1 is preserved, $\frac{1}{\sqrt{2}}$ is contractive, and if the distance $\sqrt{3}$ is extensive by a mapping $f : H_1 \to H_2$, then f is an affine isometry.*

7.3 Applications of an Inequality for Five Points

In this section, we assume that H_1 is a real (or complex) inner product space and c_1, c_2, c_3, c_4, c_5, e_1, e_2, e_3, e_4, e_5 are positive numbers such that there exist five points $x_1, x_2, x_3, x_4, x_5 \in H_1$ which satisfy the conditions in (6.8) as well as

$$\begin{aligned} &\|x_1 - x_2\| = c_1, \quad \|x_2 - x_3\| = c_2, \quad \|x_3 - x_4\| = c_3, \\ &\|x_4 - x_5\| = c_4, \quad \|x_5 - x_1\| = c_5, \\ &\|x_1 - x_3\| = e_1, \quad \|x_2 - x_4\| = e_2, \quad \|x_3 - x_5\| = e_3, \\ &\|x_4 - x_1\| = e_4, \quad \|x_5 - x_2\| = e_5, \end{aligned} \tag{7.16}$$

as we see in Fig. 7.4. (Obviously, due to (6.8), the five points x_1, x_2, x_3, x_4, x_5 lie on a two-dimensional subspace of H_1.)

The following theorem was proved in a paper [37] by S.-M. Jung and D. Nam.

Theorem 7.7 *Let H_1 and H_2 be real (or complex) inner product spaces. Assume that the distances c_1, c_2, c_3, c_4, c_5 are contractive and the distances e_1, e_2, e_3, e_4, e_5 are extensive by a mapping $f : H_1 \to H_2$, where c_i's and e_i's are given by (7.16) and the corresponding x_i's satisfy the conditions in (6.8) (see Fig. 7.4). Then f preserves all the distances c_1, c_2, c_3, c_4, c_5, e_1, e_2, e_3, e_4, and e_5.*

Proof Since the distances c_1, c_2, c_3, c_4, c_5 are contractive by f, it holds that

$$\begin{aligned} &\phi^2\big(\|x_1 - x_2\|^2 + \|x_2 - x_3\|^2 + \|x_3 - x_4\|^2 + \|x_4 - x_5\|^2 + \|x_5 - x_1\|^2\big) \\ &\geq \phi^2\big(\|f(x_1) - f(x_2)\|^2 + \|f(x_2) - f(x_3)\|^2 + \|f(x_3) - f(x_4)\|^2 \\ &\quad + \|f(x_4) - f(x_5)\|^2 + \|f(x_5) - f(x_1)\|^2\big). \end{aligned} \tag{7.17}$$

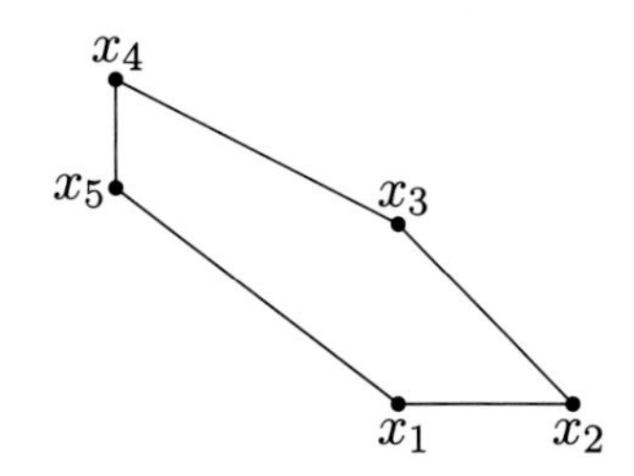

Fig. 7.4 An example is given for five points $x_1, x_2, \ldots, x_5$ that satisfy the conditions (6.8) and (7.16)

It further follows from Theorem 6.6 (i) that

$$\begin{aligned}
&\phi^2\big(\|f(x_1)-f(x_2)\|^2+\|f(x_2)-f(x_3)\|^2+\|f(x_3)-f(x_4)\|^2\\
&\quad+\|f(x_4)-f(x_5)\|^2+\|f(x_5)-f(x_1)\|^2\big)\\
&\quad\geq\|f(x_1)-f(x_3)\|^2+\|f(x_2)-f(x_4)\|^2+\|f(x_3)-f(x_5)\|^2\\
&\qquad+\|f(x_4)-f(x_1)\|^2+\|f(x_5)-f(x_2)\|^2.
\end{aligned}\tag{7.18}$$

Moreover, since the distances e_1, e_2, e_3, e_4, and e_5 are extensive by f, we consider (7.16) to obtain

$$\begin{aligned}
&\|f(x_1)-f(x_3)\|^2+\|f(x_2)-f(x_4)\|^2+\|f(x_3)-f(x_5)\|^2\\
&+\|f(x_4)-f(x_1)\|^2+\|f(x_5)-f(x_2)\|^2\\
&\quad\geq\|x_1-x_3\|^2+\|x_2-x_4\|^2+\|x_3-x_5\|^2+\|x_4-x_1\|^2\\
&\qquad+\|x_5-x_2\|^2.
\end{aligned}\tag{7.19}$$

Since x_4 and x_5 satisfy the conditions in (6.8) (see Fig. 7.4), by Theorem 6.6 (ii), we conclude that

$$\begin{aligned}
&\|x_1-x_3\|^2+\|x_2-x_4\|^2+\|x_3-x_5\|^2+\|x_4-x_1\|^2+\|x_5-x_2\|^2\\
&\quad=\phi^2\big(\|x_1-x_2\|^2+\|x_2-x_3\|^2+\|x_3-x_4\|^2+\|x_4-x_5\|^2\\
&\qquad+\|x_5-x_1\|^2\big),
\end{aligned}\tag{7.20}$$

which implies that the equality sign holds in each of (7.17), (7.18), and (7.19).

On the other hand, our hypotheses imply that

$$\begin{aligned}
c_1&=\|x_1-x_2\|\geq\|f(x_1)-f(x_2)\|,\\
c_2&=\|x_2-x_3\|\geq\|f(x_2)-f(x_3)\|,\\
c_3&=\|x_3-x_4\|\geq\|f(x_3)-f(x_4)\|,\\
c_4&=\|x_4-x_5\|\geq\|f(x_4)-f(x_5)\|,\\
c_5&=\|x_5-x_1\|\geq\|f(x_5)-f(x_1)\|,\\
e_1&=\|x_1-x_3\|\leq\|f(x_1)-f(x_3)\|,\\
e_2&=\|x_2-x_4\|\leq\|f(x_2)-f(x_4)\|,\\
e_3&=\|x_3-x_5\|\leq\|f(x_3)-f(x_5)\|,\\
e_4&=\|x_4-x_1\|\leq\|f(x_4)-f(x_1)\|,\\
e_5&=\|x_5-x_2\|\leq\|f(x_5)-f(x_2)\|.
\end{aligned}\tag{7.21}$$

By combining (7.17), (7.18), (7.19) with equality signs, (7.20), and (7.21), we conclude that

$$\begin{aligned}
\|x_1 - x_2\| &= c_1 = \|f(x_1) - f(x_2)\|,\\
\|x_2 - x_3\| &= c_2 = \|f(x_2) - f(x_3)\|,\\
\|x_3 - x_4\| &= c_3 = \|f(x_3) - f(x_4)\|,\\
\|x_4 - x_5\| &= c_4 = \|f(x_4) - f(x_5)\|,\\
\|x_5 - x_1\| &= c_5 = \|f(x_5) - f(x_1)\|,\\
\|x_1 - x_3\| &= e_1 = \|f(x_1) - f(x_3)\|,\\
\|x_2 - x_4\| &= e_2 = \|f(x_2) - f(x_4)\|,\\
\|x_3 - x_5\| &= e_3 = \|f(x_3) - f(x_5)\|,\\
\|x_4 - x_1\| &= e_4 = \|f(x_4) - f(x_1)\|,\\
\|x_5 - x_2\| &= e_5 = \|f(x_5) - f(x_2)\|.
\end{aligned}$$

For any given $x_1, x_2 \in H_1$ with $\|x_1 - x_2\| = c_1$, we can choose three points $x_3, x_4, x_5 \in H_1$ such that x_1, x_2, x_3, x_4, and x_5 determine the geometric figure that is congruent to the one in Fig. 7.4. In view of the above argument, we may conclude that $\|f(x_1) - f(x_2)\| = c_1$. For other distances such as c_2, c_3, c_4, c_5, e_1, e_2, e_3, e_4, and e_5, we can apply a similar argument. Therefore, f preserves all the distances $c_1, c_2, c_3, c_4, c_5, e_1, e_2, e_3, e_4$, and e_5. □

Remark 7.8 Assume that x_1, x_2, x_3, x_4, and x_5 are the vertices of a unit regular pentagon as we see in Fig. 7.5. If we let $c_1 = c_2 = c_3 = c_4 = c_5 = 1$ and $e_1 = e_2 = e_3 = e_4 = e_5 = \phi$ in Theorem 7.7, then we see that f preserves the distances 1 and ϕ, where ϕ is the golden ratio.

Theorem 7.9 (Jung and Nam) *Assume that H_1 and H_2 are real Hilbert spaces with* $\dim H_1 > 2$. *If the distance* 1 *is contractive and the distance ϕ is extensive by a mapping $f : H_1 \to H_2$, then f is an affine isometry.*

Proof According to Theorem 7.7 and Remark 7.8, f preserves both the distances 1 and ϕ. We claim that f preserves the distance $\sqrt{2}\phi$. Suppose v_1 and v_3 are arbitrary points of

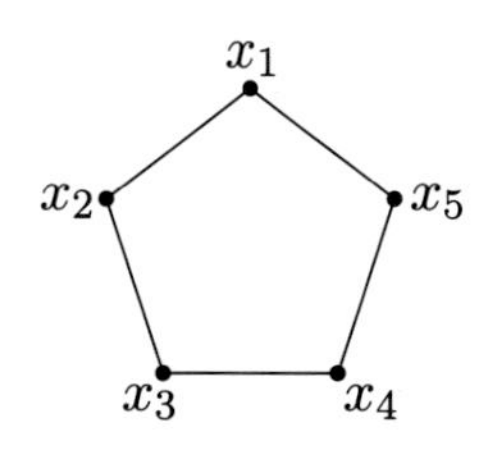

Fig. 7.5 The vertices of a unit regular pentagon are denoted as $x_1, x_2, \ldots, x_5$

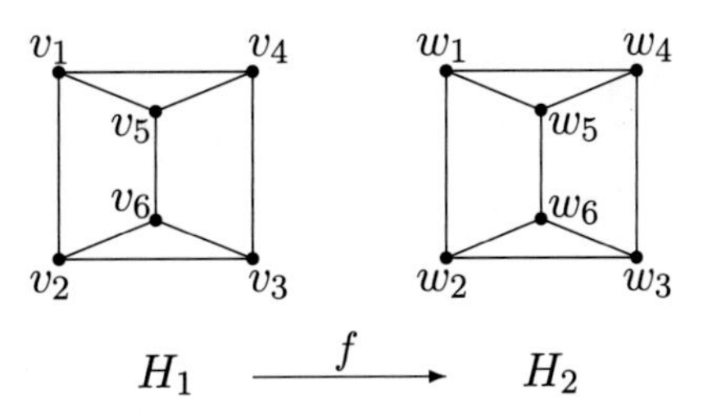

Fig. 7.6 In the left figure, v_1, v_5, v_6, v_2 are some of the vertices of a unit regular pentagon, and v_4, v_5, v_6, v_3 are some of the vertices of another unit regular pentagon

H_1 with $\|v_1 - v_3\| = \sqrt{2}\phi$. Since $\dim H_1 > 2$, there exists a subspace U of H_1 containing v_1 and v_3 that is (Hilbert space) isomorphic to 3-dimensional Euclidean space $\mathbb{E}^3$ (see Theorem 1.59).

In the first (roof-shaped) figure of Fig. 7.6, $\{v_1, v_5, v_6, v_2\}$ is a part of the vertices of a unit regular pentagon in U and also $\{v_4, v_5, v_6, v_3\}$ is a part of the vertices of another unit regular pentagon in U such that $\|v_1 - v_4\| = \|v_2 - v_3\| = \phi$. Here are the details:

$$\|v_1 - v_5\| = 1, \quad \|v_4 - v_5\| = 1, \quad \|v_5 - v_6\| = 1, \quad \|v_2 - v_6\| = 1,$$
$$\|v_3 - v_6\| = 1, \quad \|v_1 - v_2\| = \phi, \quad \|v_2 - v_3\| = \phi, \quad \|v_3 - v_4\| = \phi,$$
$$\|v_4 - v_1\| = \phi, \quad \|v_1 - v_6\| = \phi, \quad \|v_2 - v_5\| = \phi, \quad \|v_3 - v_5\| = \phi,$$
$$\|v_4 - v_6\| = \phi.$$

Now we define $w_i = f(v_i)$ for all $i \in \{1, 2, \ldots, 6\}$. Since f preserves the distances 1 and ϕ, we obtain

$$\begin{aligned}
&\|w_1 - w_5\| = 1, \quad \|w_4 - w_5\| = 1, \quad \|w_5 - w_6\| = 1, \\
&\|w_2 - w_6\| = 1, \quad \|w_3 - w_6\| = 1, \quad \|w_1 - w_2\| = \phi, \\
&\|w_2 - w_3\| = \phi, \quad \|w_3 - w_4\| = \phi, \quad \|w_4 - w_1\| = \phi, \\
&\|w_1 - w_6\| = \phi, \quad \|w_2 - w_5\| = \phi, \quad \|w_3 - w_5\| = \phi, \\
&\|w_4 - w_6\| = \phi.
\end{aligned} \tag{7.22}$$

If we set $x_i = w_i - w_1$ for $i \in \{2, 3, \ldots, 6\}$, then

$$x_j - x_k = (w_j - w_1) - (w_k - w_1) = w_j - w_k \tag{7.23}$$

for any $j, k \in \{2, 3, \ldots, 6\}$. The distances between four points w_1, w_2, w_5, and w_6 are given:

$$\begin{aligned}
\|x_2\| &= \|w_2 - w_1\| = \phi, \; \|x_5 - x_2\| = \|w_5 - w_2\| = \phi, \\
\|x_5\| &= \|w_5 - w_1\| = 1, \; \|x_6 - x_2\| = \|w_6 - w_2\| = 1, \\
\|x_6\| &= \|w_6 - w_1\| = \phi, \; \|x_6 - x_5\| = \|w_6 - w_5\| = 1.
\end{aligned} \tag{7.24}$$

Since

$$\|x_j - x_k\|^2 = \|x_j\|^2 - 2\langle x_j, x_k\rangle + \|x_k\|^2$$

for all $j, k \in \{2, 3, \ldots, 6\}$, it follows from (7.24) that

$$\langle x_2, x_5\rangle = \frac{1}{2}\left(\|x_2\|^2 + \|x_5\|^2 - \|x_2 - x_5\|^2\right) = \frac{1}{2}. \tag{7.25}$$

Similarly, we have $\langle x_2, x_6\rangle = \frac{1}{2}(2\phi^2 - 1)$ and $\langle x_5, x_6\rangle = \frac{1}{2}\phi^2$.

Hence, we can calculate $\|x_6 - x_5 - \frac{1}{\phi}x_2\|^2 = 0$ as follows:

$$\begin{aligned}
&\left\|x_6 - x_5 - \frac{1}{\phi}x_2\right\|^2 \\
&= \left\langle x_6 - x_5 - \frac{1}{\phi}x_2,\ x_6 - x_5 - \frac{1}{\phi}x_2\right\rangle \\
&= \|x_6\|^2 + \|x_5\|^2 + \frac{1}{\phi^2}\|x_2\|^2 - 2\langle x_5, x_6\rangle - \frac{2}{\phi}\langle x_2, x_6\rangle + \frac{2}{\phi}\langle x_2, x_5\rangle \\
&= \phi^2 + 1 + \frac{1}{\phi^2}\phi^2 - \phi^2 - \frac{1}{\phi}\left(2\phi^2 - 1\right) + \frac{1}{\phi} \\
&= -\frac{2}{\phi}\left(\phi^2 - \phi - 1\right) \\
&= 0.
\end{aligned}$$

Therefore, we have

$$x_6 = \frac{1}{\phi}x_2 + x_5. \tag{7.26}$$

Moreover, it follows from the definition of x_i's that

$$0 = x_6 - \frac{1}{\phi}x_2 - x_5 = w_6 - w_1 - \frac{1}{\phi}(w_2 - w_1) - (w_5 - w_1). \tag{7.27}$$

As we see in Fig. 7.6, the structures of $\{w_1, w_2, w_6, w_5\}$ and $\{w_4, w_3, w_6, w_5\}$ are congruent. Thus, we can replace w_1, w_2, w_6, w_5 in (7.27) with w_4, w_3, w_6, w_5, respectively, and use (7.23) to get

$$\begin{aligned}
0 &= w_6 - w_4 - \frac{1}{\phi}(w_3 - w_4) - (w_5 - w_4) \\
&= x_6 - x_4 - \frac{1}{\phi}(x_3 - x_4) - (x_5 - x_4).
\end{aligned}$$

Therefore, $x_6 - x_4 = \frac{1}{\phi}(x_3 - x_4) + (x_5 - x_4)$. And it follows from (7.26) and the last equality that

$$x_3 = x_2 + x_4. \tag{7.28}$$

In view of (7.22), the distances between the three points w_1, w_4, w_5 are given below.

$$\begin{aligned} \|x_4\| &= \|w_4 - w_1\| = \phi, \\ \|x_5\| &= \|w_5 - w_1\| = 1, \\ \|x_5 - x_4\| &= \|w_5 - w_4\| = 1. \end{aligned} \tag{7.29}$$

Hence, by (7.29), we obtain

$$\langle x_4, x_5 \rangle = \frac{1}{2}\left(\|x_4\|^2 + \|x_5\|^2 - \|x_4 - x_5\|^2\right) = \frac{1}{2}\phi^2. \tag{7.30}$$

Using (7.22), (7.23), (7.25), (7.26), (7.28), and (7.30), we have

$$\begin{aligned} \phi^2 &= \|w_6 - w_4\|^2 = \|x_6 - x_4\|^2 = \left\| \frac{1}{\phi} x_2 + x_5 - x_4 \right\|^2 \\ &= \frac{1}{\phi^2}\|x_2\|^2 + \|x_4\|^2 + \|x_5\|^2 + \frac{2}{\phi}\langle x_2, x_5 \rangle - \frac{2}{\phi}\langle x_2, x_4 \rangle - 2\langle x_4, x_5 \rangle \\ &= \frac{1}{\phi^2}\phi^2 + \phi^2 + 1 + \frac{1}{\phi} - \frac{2}{\phi}\langle x_2, x_4 \rangle - \phi^2 \\ &= 2 + \frac{1}{\phi} - \frac{2}{\phi}\langle x_2, x_4 \rangle. \end{aligned}$$

Hence, using the formula $\phi^2 - \phi - 1 = 0$, we get

$$\langle x_2, x_4 \rangle = \frac{\phi}{2}\left(2 + \frac{1}{\phi} - \phi^2\right) = \frac{1}{2}(\phi + 1 - \phi^2) = 0.$$

Hence, by (7.22), (7.24), (7.28) and the last equality, we get

$$\|x_3\|^2 = \|x_2 + x_4\|^2 = \|x_2\|^2 + \|x_4\|^2 = \|x_2\|^2 + \|w_4 - w_1\|^2 = 2\phi^2,$$

i.e.,

$$\|f(v_3) - f(v_1)\| = \|w_3 - w_1\| = \|x_3\| = \sqrt{2}\phi.$$

Since f preserves distances ϕ and $\sqrt{2}\phi$, we conclude that f is an affine isometry by Remark 2.19 and Theorem 5.3. □

7.4 Applications of an Inequality for n Points

In this section, we assume that H_1 is a real (or complex) inner product space and c_{12}, c_{23}, c_{34}, c_{45}, c_{56}, c_{67}, c_{71}, e_{13}, e_{24}, e_{35}, e_{46}, e_{57}, e_{61}, e_{72}, c_{14}, c_{25}, c_{36}, c_{47}, c_{51}, c_{62}, c_{73} are positive real numbers such that there exist points $x_1, x_2, x_3, x_4, x_5, x_6, x_7$ of H_1 which satisfy the condition (6.15) as well as

$$\begin{aligned}
&\|x_1 - x_2\| = c_{12}, && \|x_1 - x_3\| = e_{13}, && \|x_1 - x_4\| = c_{14},\\
&\|x_2 - x_3\| = c_{23}, && \|x_2 - x_4\| = e_{24}, && \|x_2 - x_5\| = c_{25},\\
&\|x_3 - x_4\| = c_{34}, && \|x_3 - x_5\| = e_{35}, && \|x_3 - x_6\| = c_{36},\\
&\|x_4 - x_5\| = c_{45}, && \|x_4 - x_6\| = e_{46}, && \|x_4 - x_7\| = c_{47},\\
&\|x_5 - x_6\| = c_{56}, && \|x_5 - x_7\| = e_{57}, && \|x_5 - x_1\| = c_{51},\\
&\|x_6 - x_7\| = c_{67}, && \|x_6 - x_1\| = e_{61}, && \|x_6 - x_2\| = c_{62},\\
&\|x_7 - x_1\| = c_{71}, && \|x_7 - x_2\| = e_{72}, && \|x_7 - x_3\| = c_{73},
\end{aligned} \tag{7.31}$$

as we see in Fig. 7.7. (Obviously, due to (6.15), the seven points $x_1, x_2, x_3, x_4, x_5, x_6, x_7$ lie on a two-dimensional subspace of H_1.)

The following theorem was proved in a paper [38] by S.-M. Jung and D. Nam.

Theorem 7.10 *Let H_1 and H_2 be real (or complex) inner product spaces. Assume that the distances c_{12}, c_{23}, c_{34}, c_{45}, c_{56}, c_{67}, c_{71}, c_{14}, c_{25}, c_{36}, c_{47}, c_{51}, c_{62}, c_{73} are contractive and the distances e_{13}, e_{24}, e_{35}, e_{46}, e_{57}, e_{61}, e_{72} are extensive by a mapping $f : H_1 \to H_2$, where c_{ij}'s and e_{ij}'s are given by* (7.31) *and the corresponding x_i's satisfy the condition* (6.15) *with $n = 7$ (see Fig. 7.7). Then, f preserves all the distances c_{ij}'s and e_{ij}'s.*

Proof To simplify the notations, we temporarily set $x_{i,j} = \|x_i - x_j\|$ and $y_{i,j} = \|f(x_i) - f(x_j)\|$ for all $i, j \in \{1, 2, \ldots, 7\}$ with $i \neq j$. Since the distances c_{12}, c_{23}, c_{34}, c_{45}, c_{56}, c_{67}, c_{71}, c_{14}, c_{25}, c_{36}, c_{47}, c_{51}, c_{62}, c_{73} are contractive by f, we have

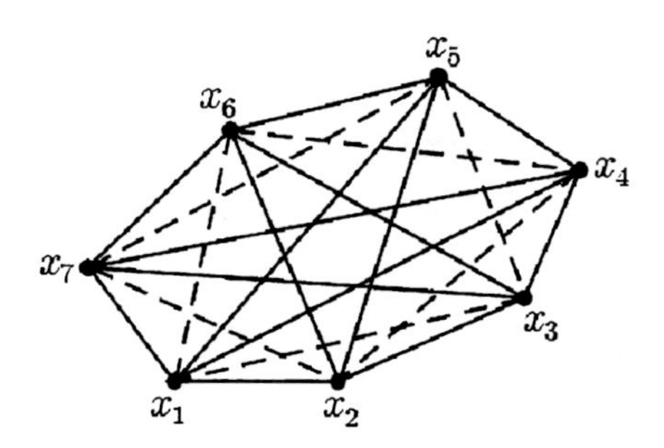

Fig. 7.7 An example is given for seven points $x_1, x_2, \ldots, x_7$ that satisfy the conditions (6.15) and (7.31)

$$
\begin{aligned}
&(c_7^2 + 2c_7)(x_{1,2}^2 + x_{2,3}^2 + x_{3,4}^2 + x_{4,5}^2 + x_{5,6}^2 + x_{6,7}^2 + x_{7,1}^2)\\
&\quad + x_{1,4}^2 + x_{2,5}^2 + x_{3,6}^2 + x_{4,7}^2 + x_{5,1}^2 + x_{6,2}^2 + x_{7,3}^2\\
&\geq (c_7^2 + 2c_7)(y_{1,2}^2 + y_{2,3}^2 + y_{3,4}^2 + y_{4,5}^2 + y_{5,6}^2 + y_{6,7}^2 + y_{7,1}^2)\\
&\quad + y_{1,4}^2 + y_{2,5}^2 + y_{3,6}^2 + y_{4,7}^2 + y_{5,1}^2 + y_{6,2}^2 + y_{7,3}^2.
\end{aligned}
\tag{7.32}
$$

Now we can use Theorem 6.9 (i) to get

$$
\begin{aligned}
&(c_7^2 + 2c_7)(y_{1,2}^2 + y_{2,3}^2 + y_{3,4}^2 + y_{4,5}^2 + y_{5,6}^2 + y_{6,7}^2 + y_{7,1}^2)\\
&\quad + y_{1,4}^2 + y_{2,5}^2 + y_{3,6}^2 + y_{4,7}^2 + y_{5,1}^2 + y_{6,2}^2 + y_{7,3}^2\\
&\geq 2c_7(y_{1,3}^2 + y_{2,4}^2 + y_{3,5}^2 + y_{4,6}^2 + y_{5,7}^2 + y_{6,1}^2 + y_{7,2}^2).
\end{aligned}
\tag{7.33}
$$

Since the distances $e_{13}, e_{24}, e_{35}, e_{46}, e_{57}, e_{61}, e_{72}$ are extensive by f, we obtain

$$
\begin{aligned}
&2c_7(y_{1,3}^2 + y_{2,4}^2 + y_{3,5}^2 + y_{4,6}^2 + y_{5,7}^2 + y_{6,1}^2 + y_{7,2}^2)\\
&\geq 2c_7(x_{1,3}^2 + x_{2,4}^2 + x_{3,5}^2 + x_{4,6}^2 + x_{5,7}^2 + x_{6,1}^2 + x_{7,2}^2).
\end{aligned}
\tag{7.34}
$$

Since the x_i's satisfy the conditions in (6.15), it follows from Theorem 6.9 (ii) that

$$
\begin{aligned}
&2c_7(x_{1,3}^2 + x_{2,4}^2 + x_{3,5}^2 + x_{4,6}^2 + x_{5,7}^2 + x_{6,1}^2 + x_{7,2}^2)\\
&= (c_7^2 + 2c_7)(x_{1,2}^2 + x_{2,3}^2 + x_{3,4}^2 + x_{4,5}^2 + x_{5,6}^2 + x_{6,7}^2 + x_{7,1}^2)\\
&\quad + x_{1,4}^2 + x_{2,5}^2 + x_{3,6}^2 + x_{4,7}^2 + x_{5,1}^2 + x_{6,2}^2 + x_{7,3}^2.
\end{aligned}
\tag{7.35}
$$

Altogether, we arrive at the conclusion that the equal sign must hold in each of the inequalities (7.32), (7.33), (7.34), and (7.35). In particular, it holds that

$$
\begin{aligned}
&(c_7^2 + 2c_7)(x_{1,2}^2 + x_{2,3}^2 + x_{3,4}^2 + x_{4,5}^2 + x_{5,6}^2 + x_{6,7}^2 + x_{7,1}^2)\\
&\quad + x_{1,4}^2 + x_{2,5}^2 + x_{3,6}^2 + x_{4,7}^2 + x_{5,1}^2 + x_{6,2}^2 + x_{7,3}^2\\
&= (c_7^2 + 2c_7)(y_{1,2}^2 + y_{2,3}^2 + y_{3,4}^2 + y_{4,5}^2 + y_{5,6}^2 + y_{6,7}^2 + y_{7,1}^2)\\
&\quad + y_{1,4}^2 + y_{2,5}^2 + y_{3,6}^2 + y_{4,7}^2 + y_{5,1}^2 + y_{6,2}^2 + y_{7,3}^2
\end{aligned}
\tag{7.36}
$$

and

$$
\begin{aligned}
&x_{1,3}^2 + x_{2,4}^2 + x_{3,5}^2 + x_{4,6}^2 + x_{5,7}^2 + x_{6,1}^2 + x_{7,2}^2\\
&\quad = y_{1,3}^2 + y_{2,4}^2 + y_{3,5}^2 + y_{4,6}^2 + y_{5,7}^2 + y_{6,1}^2 + y_{7,2}^2.
\end{aligned}
\tag{7.37}
$$

On the other hand, our hypotheses imply that

$$
\begin{array}{lll}
c_{12} = x_{1,2} \geq y_{1,2}, & c_{14} = x_{1,4} \geq y_{1,4}, & c_{23} = x_{2,3} \geq y_{2,3}, \\
c_{25} = x_{2,5} \geq y_{2,5}, & c_{34} = x_{3,4} \geq y_{3,4}, & c_{36} = x_{3,6} \geq y_{3,6}, \\
c_{45} = x_{4,5} \geq y_{4,5}, & c_{47} = x_{4,7} \geq y_{4,7}, & c_{56} = x_{5,6} \geq y_{5,6}, \\
c_{51} = x_{5,1} \geq y_{5,1}, & c_{67} = x_{6,7} \geq y_{6,7}, & c_{62} = x_{6,2} \geq y_{6,2}, \\
c_{71} = x_{7,1} \geq y_{7,1}, & c_{73} = x_{7,3} \geq y_{7,3}, & e_{13} = x_{1,3} \leq y_{1,3}, \\
e_{24} = x_{2,4} \leq y_{2,4}, & e_{35} = x_{3,5} \leq y_{3,5}, & e_{46} = x_{4,6} \leq y_{4,6}, \\
e_{57} = x_{5,7} \leq y_{5,7}, & e_{61} = x_{6,1} \leq y_{6,1}, & e_{72} = x_{7,2} \leq y_{7,2}.
\end{array}
\tag{7.38}
$$

By combining (7.36), (7.37), and (7.38), we conclude that

$$
\begin{array}{lll}
x_{1,2} = c_{12} = y_{1,2}, & x_{1,4} = c_{14} = y_{1,4}, & x_{2,3} = c_{23} = y_{2,3}, \\
x_{2,5} = c_{25} = y_{2,5}, & x_{3,4} = c_{34} = y_{3,4}, & x_{3,6} = c_{36} = y_{3,6}, \\
x_{4,5} = c_{45} = y_{4,5}, & x_{4,7} = c_{47} = y_{4,7}, & x_{5,6} = c_{56} = y_{5,6}, \\
x_{5,1} = c_{51} = y_{5,1}, & x_{6,7} = c_{67} = y_{6,7}, & x_{6,2} = c_{62} = y_{6,2}, \\
x_{7,1} = c_{71} = y_{7,1}, & x_{7,3} = c_{73} = y_{7,3}, & x_{1,3} = e_{13} = y_{1,3}, \\
x_{2,4} = e_{24} = y_{2,4}, & x_{3,5} = e_{35} = y_{3,5}, & x_{4,6} = e_{46} = y_{4,6}, \\
x_{5,7} = e_{57} = y_{5,7}, & x_{6,1} = e_{61} = y_{6,1}, & x_{7,2} = e_{72} = y_{7,2}.
\end{array}
$$

For example, with (7.38), we can check that $x_{1,2} = c_{12} = y_{1,2}$: If $x_{1,2} > y_{1,2}$, then it follows from (7.38) that

$$
\begin{aligned}
&\left(c_7^2 + 2c_7\right)\left(x_{1,2}^2 + x_{2,3}^2 + x_{3,4}^2 + x_{4,5}^2 + x_{5,6}^2 + x_{6,7}^2 + x_{7,1}^2\right) \\
&\quad + x_{1,4}^2 + x_{2,5}^2 + x_{3,6}^2 + x_{4,7}^2 + x_{5,1}^2 + x_{6,2}^2 + x_{7,3}^2 \\
&> \left(c_7^2 + 2c_7\right)\left(y_{1,2}^2 + y_{2,3}^2 + y_{3,4}^2 + y_{4,5}^2 + y_{5,6}^2 + y_{6,7}^2 + y_{7,1}^2\right) \\
&\quad + y_{1,4}^2 + y_{2,5}^2 + y_{3,6}^2 + y_{4,7}^2 + y_{5,1}^2 + y_{6,2}^2 + y_{7,3}^2,
\end{aligned}
$$

which contradicts (7.36). Thus, considering (7.38), we conclude that $x_{1,2} = y_{1,2}$.

For any given $x_1, x_2 \in H_1$ with $\|x_1 - x_2\| = x_{1,2} = c_{12}$, we can choose five points x_3, x_4, x_5, x_6, x_7 in H_1 such that $x_1, x_2, \ldots, x_7$ determine a geometric figure congruent to that shown in Fig. 7.7. In view of the above argument, we may conclude that $\|f(x_1) - f(x_2)\| = y_{1,2} = c_{12}$. For other distances such as $c_{23}, c_{34}, c_{45}, c_{56}, c_{67}, c_{71}, c_{14}, c_{25}, c_{36}, c_{47}, c_{51}, c_{62}, c_{73}, e_{13}, e_{24}, e_{35}, e_{46}, e_{57}, e_{61}$, and e_{72}, we can apply a similar argument. Therefore, f preserves the distances $c_{12}, c_{23}, c_{34}, c_{45}, c_{56}, c_{67}, c_{71}, c_{14}, c_{25}, c_{36}, c_{47}, c_{51}, c_{62}, c_{73}, e_{13}, e_{24}, e_{35}, e_{46}, e_{57}, e_{61}$, and e_{72}. □

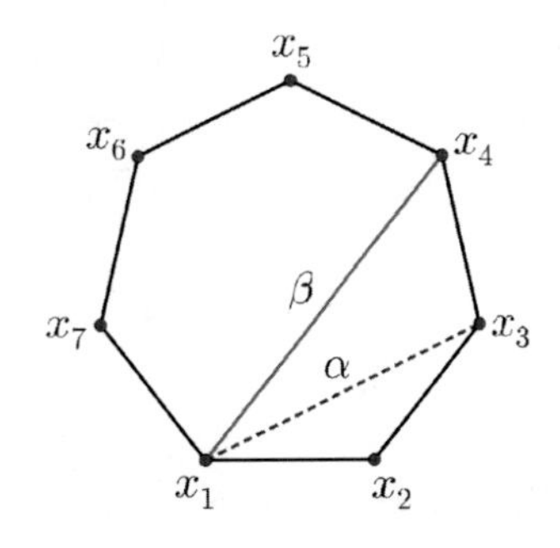

Fig. 7.8 The seven points $x_1, x_2, \ldots, x_7$ are the vertices of a unit regular heptagon whose diagonal lengths are α and β

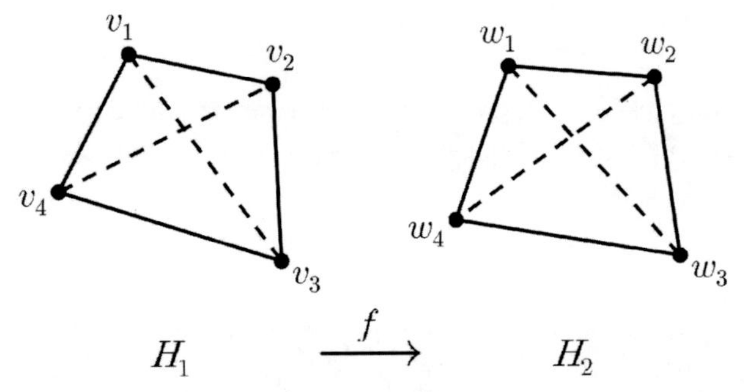

Fig. 7.9 The four points v_1, v_2, v_3, v_4 lie on a plane. Note that $w_i = f(v_i)$ for all $i \in \{1, 2, 3, 4\}$

Remark 7.11 Assume that x_1, x_2, x_3, x_4, x_5, x_6, x_7 are the vertices of a unit regular heptagon S (see Fig. 7.8). Let $\alpha = \frac{\sin \frac{2\pi}{7}}{\sin \frac{\pi}{7}} \approx 1.8019\ldots$ be the shorter diagonal and $\beta = c_7 = \frac{\sin \frac{3\pi}{7}}{\sin \frac{\pi}{7}} \approx 2.2469\ldots$ be the longer diagonal of S. If we set $c_{12} = c_{23} = c_{34} = c_{45} = c_{56} = c_{67} = c_{71} = 1$, $e_{13} = e_{24} = e_{35} = e_{46} = e_{57} = e_{61} = e_{72} = \alpha$ and $c_{14} = c_{25} = c_{36} = c_{47} = c_{51} = c_{62} = c_{73} = \beta$ in Theorem 7.10, then the mapping f given in Theorem 7.10 preserves the distances 1, α, and β.

Lemma 7.12 *Assume that H_1 and H_2 are real inner product spaces, and let $f : H_1 \to H_2$ be a mapping. Let v_1, v_2, v_3, v_4 be arbitrary four points in H_1. If v_1, v_2, v_3, v_4 lie on one plane and $\|v_i - v_j\| = \|f(v_i) - f(v_j)\|$ for all $i, j \in \{1, 2, 3, 4\}$ with $i < j$, then $f(v_1)$, $f(v_2)$, $f(v_3)$, $f(v_4)$ also lie on one plane.*

Proof In Fig. 7.9, each w_i stands for $f(v_i)$. Since the translation preserves distances between points and does not affect coplanarity of points, we assume that $v_4 = f(v_4) = 0$ without loss of generality. Then the condition of this lemma becomes simple as follows:

$$\begin{aligned} \|v_1\| &= \|f(v_1)\|, \quad \|v_1 - v_2\| = \|f(v_1) - f(v_2)\|, \\ \|v_2\| &= \|f(v_2)\|, \quad \|v_2 - v_3\| = \|f(v_2) - f(v_3)\|, \\ \|v_3\| &= \|f(v_3)\|, \quad \|v_3 - v_1\| = \|f(v_3) - f(v_1)\|. \end{aligned}$$

It follows from the above condition that

$$\begin{aligned}\langle v_1, v_2\rangle &= \frac{1}{2}\big(\|v_1\|^2 + \|v_2\|^2 - \|v_1 - v_2\|^2\big)\\ &= \frac{1}{2}\big(\|f(v_1)\|^2 + \|f(v_2)\|^2 - \|f(v_1) - f(v_2)\|^2\big)\\ &= \langle f(v_1), f(v_2)\rangle.\end{aligned}$$

By a similar way, we obtain

$$\langle v_2, v_3\rangle = \langle f(v_2), f(v_3)\rangle \quad \text{and} \quad \langle v_3, v_1\rangle = \langle f(v_3), f(v_1)\rangle.$$

Because $v_1, v_2, v_3, v_4 (= 0)$ lie on one plane, i.e., they are coplanar, there exists $r_1, r_2 \in \mathbb{R}$ satisfying $v_3 = r_1 v_1 + r_2 v_2$. Hence, $0 = v_3 - r_1 v_1 - r_2 v_2$, and so

$$\begin{aligned}0 &= \|v_3 - r_1 v_1 - r_2 v_2\|^2\\ &= \|v_3\|^2 + r_1^2\|v_1\|^2 + r_2^2\|v_2\|^2\\ &\quad - 2r_1\langle v_3, v_1\rangle - 2r_2\langle v_3, v_2\rangle + 2r_1 r_2\langle v_1, v_2\rangle\\ &= \|f(v_3)\|^2 + r_1^2\|f(v_1)\|^2 + r_2^2\|f(v_2)\|^2\\ &\quad - 2r_1\langle f(v_3), f(v_1)\rangle - 2r_2\langle f(v_3), f(v_2)\rangle + 2r_1 r_2\langle f(v_1), f(v_2)\rangle\\ &= \|f(v_3) - r_1 f(v_1) - r_2 f(v_2)\|^2,\end{aligned}$$

i.e., $f(v_3) - r_1 f(v_1) - r_2 f(v_2) = 0$. Therefore, $f(v_3) = r_1 f(v_1) + r_2 f(v_2)$, and $f(v_1)$, $f(v_2)$, $f(v_3)$, $f(v_4)(= 0)$ also lie on one plane. □

We define α and β in the following theorem exactly as in Remark 7.11, i.e., $\alpha = \frac{\sin\frac{2\pi}{7}}{\sin\frac{\pi}{7}} \approx 1.8019\ldots$ and $\beta = \frac{\sin\frac{3\pi}{7}}{\sin\frac{\pi}{7}} \approx 2.2469\ldots$

Theorem 7.13 (Jung and Nam) *Assume that H_1 and H_2 are real Hilbert spaces with $\dim H_1 > 2$. If the distances 1 and α are contractive and the distance β is extensive by a mapping $f : H_1 \to H_2$, then f is an affine isometry.*

Proof According to Theorem 7.10 and Remark 7.11, the mapping f preserves the three distances 1, α, and β.

We claim that f preserves the distance $\sqrt{2}$. Assume that the distance between two points v_1 and v_3 of H_1 is $\sqrt{2}$, i.e., $\|v_1 - v_3\| = \sqrt{2}$. Since $\dim H_1 > 2$, there exists a 3-dimensional subspace U of H_1 that contains v_1 and v_3. Due to Theorem 1.59, there is a Hilbert space isomorphism between two finite-dimensional Hilbert spaces if they have the same dimension. Since $\dim U = 3 = \dim \mathbb{E}^3$, there is a Hilbert space isomorphism

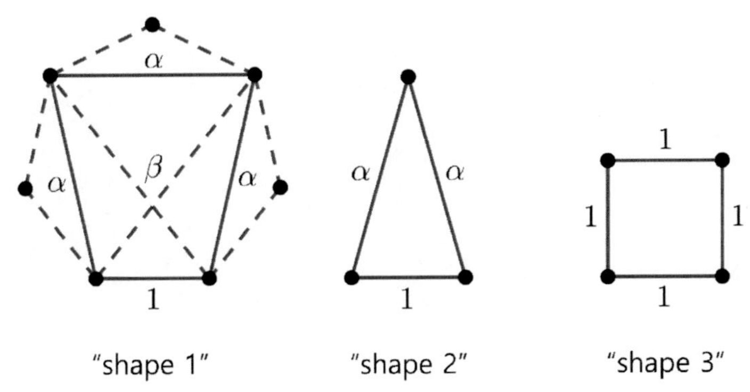

Fig. 7.10 The "shape 1" represents an isosceles trapezoid, "shape 2" represents an isosceles triangle, and "shape 3" represents a unit square

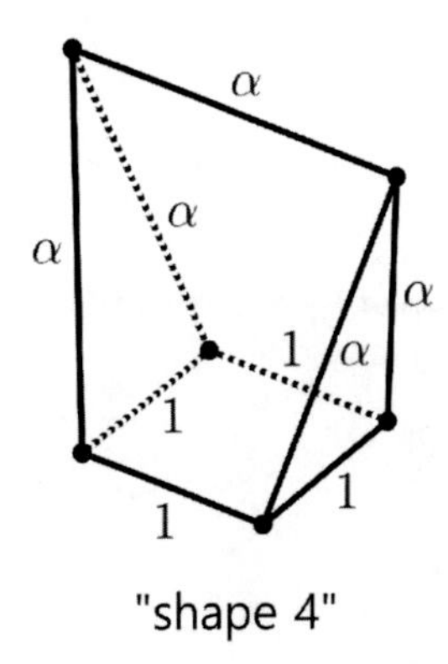

Fig. 7.11 "shape 4" is an assembly of two "shape 1s," two "shape 2s," and a "shape 3"

$L_1 : \mathbb{E}^3 \to U$. As shown in Fig. 7.10, we can draw an isosceles trapezoid on a unit regular heptagon in $\mathbb{E}^3$.

Let "shape 1" denote the isosceles trapezoid illustrated in Fig. 7.10, "shape 2" denote a triangle whose side lengths are $1, \alpha, \alpha$, and let "shape 3" denote a unit square. In $\mathbb{E}^3$, two "shape 1s," two "shape 2s," and one "shape 3" are assembled into a geometric figure, as shown in Fig. 7.11. Let "shape 4" denote this figure.

There are two points $u_1, u_3 \in \mathbb{E}^3$ such that $L_1(u_1) = v_1$ and $L_1(u_3) = v_3$. Since the Hilbert space isomorphism preserves distance, it holds that $\|u_1 - u_3\| = \|L_1(u_1) - L_1(u_3)\| = \|v_1 - v_3\| = \sqrt{2}$. Thus, we can choose u_2, u_4, u_5, and u_6 in $\mathbb{E}^3$, so that the pair of six points $\{u_1, u_2, \ldots, u_6\}$ comprises the vertices of "shape 4" (Fig. 7.11). Let $v_i = L_1(u_i)$ and $w_i = f(v_i)$ for each $i \in \{1, 2, \ldots, 6\}$. (See Fig. 7.12.)

We note that $\{u_1, u_2, u_5, u_6\}$ and $\{u_3, u_4, u_5, u_6\}$ are coplanar, respectively. Since both L_1 and f preserve the distances 1, α, and β, we can conclude that each of $\{v_1, v_2, v_5, v_6\}$ and $\{v_3, v_4, v_5, v_6\}$ is coplanar. In view of Lemma 7.12, we also conclude that each of $\{w_1, w_2, w_5, w_6\}$ and $\{w_3, w_4, w_5, w_6\}$ is coplanar. All of them compose "shape 1." We will now prove that $\|w_1 - w_3\| = \sqrt{2}$.

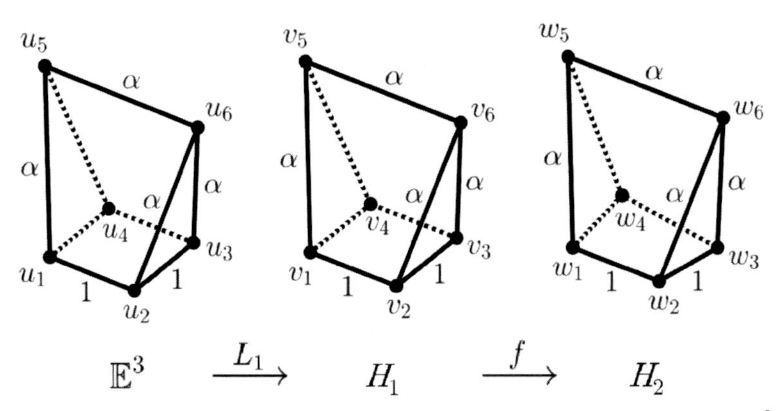

Fig. 7.12 Note that $v_i = L_1(u_i)$ and $w_i = f(v_i)$ for all $i \in \{1, 2, \ldots, 6\}$, where $L_1 : \mathbb{E}^3 \to U$ is a Hilbert space isomorphism

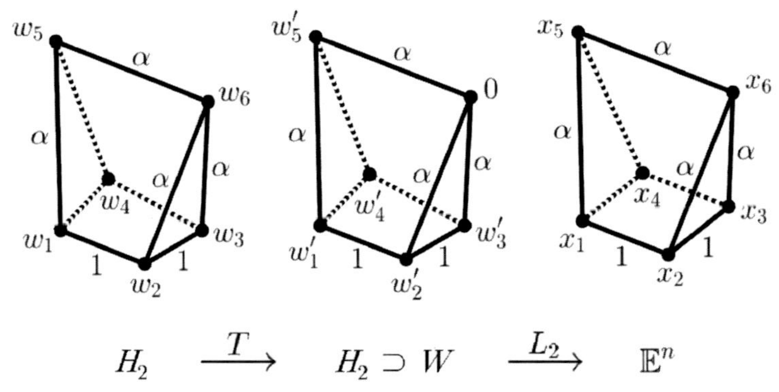

Fig. 7.13 Note that $T : H_2 \to H_2$ is a translation and $L_2 : W \to \mathbb{E}^n$ is a Hilbert space isomorphism

We define $T : H_2 \to H_2$ by $T(x) = x - w_6$, and we set $w_i' = T(w_i) = w_i - w_6$ for every $i \in \{1, 2, \ldots, 5\}$. Since T is a translation, T preserves all distances. By Lemma 7.12, each of $\{0, w_1', w_2', w_5'\}$ and $\{0, w_3', w_4', w_5'\}$ is coplanar, and all of them compose "shape 1." Thus, w_1' is a linear combination of w_2' and w_5', and w_4' is a linear combination of w_3' and w_5'. Let W be a subspace of H_2 spanned by $\{w_2', w_3', w_5'\}$. Then, $\{0, w_1', \ldots, w_5'\} \subset W$ and $\dim W \leq 3$. Thus, there exists a Hilbert space isomorphism $L_2 : W \to \mathbb{E}^n$ with $n \leq 3$ (see Fig. 7.13). We note that L_2 preserves all distances.

Let $L_2(w_i') = x_i$ for all $i \in \{1, 2, \ldots, 5\}$ and $L_2(0) = x_6$. Then each of $\{x_1, x_2, x_5, x_6\}$ and $\{x_3, x_4, x_5, x_6\}$ is coplanar by Lemma 7.12. Hence, each of $\{x_1, x_2, x_5, x_6\}$ and $\{x_3, x_4, x_5, x_6\}$ composes "shape 1." In addition, each of $\{x_1, x_4, x_5\}$ and $\{x_2, x_3, x_6\}$ composes "shape 2," and we know $\|x_2 - x_3\| = \|x_1 - x_4\| = 1$. For such a structure to be possible, n must be at least 3. Thus, we conclude that $n = 3$.

Since the points $x_1, x_2, \ldots, x_6$ are in $\mathbb{E}^3$ with the property that each pair of $\{x_1, x_2, x_5, x_6\}$ and $\{x_3, x_4, x_5, x_6\}$ composes "shape 1," and each of $\{x_1, x_4, x_5\}$ and $\{x_2, x_3, x_6\}$ composes "shape 2," we conclude that $\{x_1, x_2, x_3, x_4\}$ should compose "shape 3." Hence, we have $\|x_1 - x_3\| = \sqrt{2}$. Since T and L_2 preserve all distances, it follows that $\|w_1 - w_3\| = \sqrt{2}$. Therefore, f preserves the distance $\sqrt{2}$. Since f preserves distances 1 and $\sqrt{2}$, it follows from Theorem 5.3 that f is an affine isometry. □

Miscellaneous

8

Abstract

The Beckman-Quarles theorem states that every unit-distance preserving mapping $f : \mathbb{E}^n \to \mathbb{E}^n$ is an isometry if n is an integer greater than 1. Section 8.1 is devoted to the discussion of whether the Beckman-Quarles theorem also holds in rational n-spaces. It is known that every unit-distance preserving mapping $f : \mathbb{Q}^n \to \mathbb{Q}^n$ is an isometry if n is an even integer greater than 5 or 5 or an odd integer of the form $n = 2m^2 - 1$, where $m > 2$. We have to omit the interesting proofs of all theorems introduced in this section due to space constraints. In Sect. 8.2, we will discuss the theory of tensegrity structures that F. Rádo et al. used to partially solve the Aleksandrov-Rassias problems. Most of the content in this section comes from the paper by Bezdek and Connelly (Period Math Hungar 39(1–3):185–200, 1999). Indeed, they were able to improve the result of Rádo et al. even further by refining the idea presented by Rádo et al. In Sects. 8.3 and 8.4, we provide some sufficient conditions for the Benz-Berens theorem and the Beckman-Quarles theorem to also hold in an open convex set. The contents of those sections are mainly based on the papers by Jung (Bull Braz Math Soc (NS) 37(3):351–359, 2006); Jung (Bull Braz Math Soc (NS)40(1):77–84, 2009); Jung and Rassias (J Korean Math Soc 41(4):667–680, 2004). In the final section, we assume that the Beckman-Quarles theorem does not assume that the mapping preserves a certain distance but rather a certain geometric figure. S.-M. Jung and B. Kim have achieved interesting results on this topic, which we will systematically present in the last section.

8.1 Discrete Versions of Theorem of Beckman and Quarles

As we have seen in Sect. 2.3, Aleksandrov asked in 1970 whether the existence of a single conservative distance for a mapping implies that the mapping is an isometry. However, it is

S.-M. Jung, *Aleksandrov-Rassias Problems on Distance Preserving Mappings*, Frontiers in Mathematics, https://doi.org/10.1007/978-3-031-77613-7_8

interesting to note that Beckman and Quarles [3] partially solved the Aleksandrov problem by presenting the following theorem in 1953, long before the Aleksandrov problem was even posed.

Let n be a fixed integer greater than 1. If a mapping $f : \mathbb{R}^n \to \mathbb{R}^n$ preserves the unit distance, then f is an affine isometry.

This famous theorem has prompted a number of mathematicians to turn their attention to research in this area and has also inspired the author to write this book.

Let $\mathbb{Q}^n$ denote the rational n-dimensional space (rational n-space) equipped with the usual Euclidean metric. The following theorem presented by A. Tyszka [67] can be considered as a discrete version of the Beckman-Quarles theorem for $n = 2$.

For all points $x = (x_1, x_2)$ and $y = (y_1, y_2)$ in $\mathbb{R}^2$, we use the notation $\|x - y\|$ to denote the distance between x and y, i.e., $\|x - y\| = \sqrt{(x_1 - y_1)^2 + (x_2 - y_2)^2}$.

Theorem 8.1 (Tyszka) *If $x, y \in \mathbb{R}^2$ and $\|x - y\|$ can be constructed with a ruler and compass, then there is a finite subset S_{xy} of $\mathbb{R}^2$ with the following properties:*

(i) $x, y \in S_{xy}$;
(ii) *Every unit-distance preserving mapping $f : S_{xy} \to \mathbb{R}^2$ preserves distance between x and y.*

In addition to classifying real numbers into rational and irrational numbers, there is another way to classify real numbers into algebraic and transcendental numbers.

Definition 8.2 If a real number is a solution to a nonzero polynomial equation of the form

$$c_n x^n + c_{n-1} x^{n-1} + \cdots + c_1 x + c_0 = 0$$

with integral coefficients, the real number is said to be an *algebraic number*. When a real number is not a solution to every nonzero polynomial equation of the form mentioned above, it is said to be a *transcendental number*.

Complex numbers are also classified into algebraic and transcendental numbers in exactly the same way, but we will only focus on real numbers here.

By D_2, we denote the set of all real numbers $\rho \geq 0$ such that for every pair of two points $x, y \in \mathbb{R}^2$ with $\|x - y\| = \rho$, there is a finite subset S_{xy} of $\mathbb{R}^2$ with the properties (i) and (ii) presented in Theorem 8.1. We then remark that $\rho \in D_2$ if and only if ρ is an *algebraic number* (see [67]).

Furthermore, Tyszka was able to extend Theorem 8.1 to the case of n-dimensional Euclidean spaces in his paper [68].

Theorem 8.3 (Tyszka) *Let n be an integer greater than 1. If $x, y \in \mathbb{R}^n$ and $\|x - y\|$ is an algebraic number, then there is a finite subset S_{xy} of $\mathbb{R}^n$ with the following properties:*

(i) $x, y \in S_{xy}$;
(ii) Every unit-distance preserving mapping $f : S_{xy} \to \mathbb{R}^n$ preserves distance between x and y.

Similar to the case of Theorem 8.1, we denote by D_n the set of all real numbers $\rho \geq 0$ such that for every pair of two points $x, y \in \mathbb{R}^n$ with $\|x - y\| = \rho$, there is a finite subset S_{xy} of $\mathbb{R}^n$ with the properties (i) and (ii) presented in Theorem 8.3. We note that the set D_n plays a key role in the proof of Theorem 8.3.

Tyszka [69] succeeded in proving not only these theorems but also the following theorems.

Theorem 8.4 (Tyszka) *Let n be an integer greater than 1. If $x, y \in \mathbb{R}^n$ and $\|x - y\|$ are constructible by means of ruler and compass, then there is a finite subset S_{xy} of $\mathbb{R}^n$ with the following properties:*

(i) $x, y \in S_{xy}$;
(ii) Every unit-distance preserving mapping $f : S_{xy} \to \mathbb{R}^n$ preserves distance between x and y.

The set D_n also plays a key role in the proof of Theorem 8.4.

A subset $\mathbb{F}$ of $\mathbb{R}$ is called a *Euclidean field* if and only if for all $x \in \mathbb{F}$ there exists a $y \in \mathbb{F}$ such that $x = y^2$ or $x = -y^2$. The real constructible numbers, those (signed) lengths that can be constructed from a rational segment using ruler and compass constructions, form a Euclidean field.

Remark 8.5 Let n be an integer greater than 1. The proof of Theorem 8.4 implies that if $x, y \in \mathbb{F}^n$ and $\|x - y\|$ are constructible by means of ruler and compass, then there is a finite subset S_{xy} of $\mathbb{F}^n$ that contains x and y such that every unit-distance preserving mapping $f : S_{xy} \to \mathbb{R}^n$ preserves the distance between x and y.

Theorem 8.6 (Tyszka) *If $x, y \in \mathbb{Q}^8$, then there exists a finite subset S_{xy} of $\mathbb{Q}^8$ with the following properties:*

(i) $x, y \in S_{xy}$;
(ii) Every unit-distance preserving mapping $f : S_{xy} \to \mathbb{R}^8$ preserves distance between x and y.

Theorem 8.6 (ii) implies that any unit-distance preserving mapping from $\mathbb{Q}^8$ to $\mathbb{Q}^8$ must preserve the distance between any two points of $\mathbb{Q}^8$. Therefore, Theorem 8.6 implies the validity of the following corollary.

Corollary 8.7 (Tyszka) *Any unit-distance preserving mapping* $f : \mathbb{Q}^8 \to \mathbb{Q}^8$ *is an isometry.*

Another statement that corresponds to Corollary 8.7 is: Any mapping $f : \mathbb{Q}^8 \to \mathbb{Q}^8$ that satisfies (DOPP) is an isometry.

Tyszka [69] asked whether there are other values of n than 8 for which Theorem 8.6 and Corollary 8.7 hold. J. Zaks [74] immediately replied to Tyszka's question as follows:

Theorem 8.8 (Zaks) *If n is an even number of the form $n = 4m(m+1)$, where $m \in \mathbb{N}$, and if x, y are any two points of $\mathbb{Q}^n$, then there exists a finite subset S_{xy} of $\mathbb{Q}^n$ such that every unit-distance preserving mapping $f : S_{xy} \to \mathbb{Q}^n$ preserves the distance between x and y.*

Theorem 8.9 (Zaks) *If n is a perfect square of the form $n = 2m^2 - 1$ for some $m > 1$ and if x, y are any two points of $\mathbb{Q}^n$, then there exists a finite subset S_{xy} of $\mathbb{Q}^n$ containing x and y such that every unit-distance preserving mapping $f : S_{xy} \to \mathbb{Q}^n$ preserves the distance between x and y.*

Corollary 8.10 (Zaks) *If n is an even number of the form $n = 4m(m+1)$ for some $m \in \mathbb{N}$, then every unit-distance preserving mapping $f : \mathbb{Q}^n \to \mathbb{Q}^n$ is an isometry.*

Corollary 8.11 (Zaks) *If n is a perfect square of the form $n = 2m^2 - 1$ for some $m > 1$, then every unit-distance preserving mapping $f : \mathbb{Q}^n \to \mathbb{Q}^n$ is an isometry.*

Additionally, Zaks presented the following conjecture:

For every integer $n > 4$ and for any pair of two points $x, y \in \mathbb{Q}^n$, there is a finite subset S_{xy} of $\mathbb{Q}^n$ with the following properties:

- (i) $x, y \in S_{xy}$;
- (ii) *Every unit-distance preserving mapping $f : S_{xy} \to \mathbb{Q}^n$ preserves distance between x and y.*

If the above conjecture is correct, it is reinterpreted as an interesting conjecture as follows:

For every integer $n > 4$, every unit-distance preserving mapping $f : \mathbb{Q}^n \to \mathbb{Q}^n$ is an isometry.

For convenience, we will call this conjecture the *conjecture of Tyszka and Zaks*.

We define the subset $\tilde{N}$ of $\mathbb{N}$ by

$$\tilde{N} = \{4m(m+1) : m \in \mathbb{N}\} \cup \{2m^2 - 1 : m \in \mathbb{N},\ m > 2\}.$$

By proving the following theorem, Zaks [75] has showed that the Tyszka-Zaks conjecture is partially true.

Theorem 8.12 (Zaks) *If $n \in \tilde{N}$, then each unit-distance preserving mapping $f : \mathbb{Q}^n \to \mathbb{Q}^n$ is an isometry.*

Remark 8.13 R. Connelly and J. Zaks [15] proved that for all even integers $n > 5$, every unit-distance preserving mapping $f : \mathbb{Q}^n \to \mathbb{Q}^n$ is necessarily an isometry. In addition, W. Hibi [22–24] proved that every unit-distance preserving mapping $f : \mathbb{Q}^n \to \mathbb{Q}^n$ is an isometry, if n is an integer greater than or equal to 5. On the other hand, W. Benz [5, 6] and H. Lenz [44] proved that in the case of $n \in \{2, 3, 4\}$, there is a mapping $f : \mathbb{Q}^n \to \mathbb{Q}^n$ that preserves unit distance but is not an isometry.

Moreover, Benz [5, 6] proved that every mapping $f : \mathbb{Q}^n \to \mathbb{Q}^n$ that preserves the distances 1 and 2 is an isometry if n is greater than 4. On the other hand, Zaks [76] proved that every mapping $f : \mathbb{Q}^n \to \mathbb{Q}^n$ that preserves the distances 1 and $\sqrt{2}$ is an isometry, assuming $n > 4$.

According to [19], it is also known that every injective mapping $f : \mathbb{Q}^n \to \mathbb{Q}^n$, which preserves the distances $\frac{1}{2}\rho$ and $\rho > 0$, where ρ is rational, is an isometry if $n > 4$. From [44], it follows that every mapping $f : \mathbb{Q}^n \to \mathbb{Q}^n$ $(n > 4)$ that preserves the distances 1 and 4 is an isometry. On the other hand, by Farrahi [18, 19], we may conclude that for any $n \in \{1, 2, 3, 4\}$, there is a bijective mapping $f : \mathbb{Q}^n \to \mathbb{Q}^n$ that preserves all distances $\rho \in \{\frac{1}{2}k : k \in \mathbb{N}\}$ but is not an isometry.

As we have seen in this section, especially in Theorem 8.12 and Remark 8.13, it is not yet entirely clear whether the Tyszka-Zaks conjecture is true. It might be worth proving whether this conjecture is true or false.

8.2 Remarks on the Aleksandrov-Rassias Problems

Let n be an integer greater than 1. Assume that $p = (p_1, p_2, \ldots, p_k)$ is a *configuration* of k labeled points of $\mathbb{E}^n$. Furthermore, assume that G is a graph without loops or multiple edges, whose vertices are those k points and whose edges are specified by *cables* or *struts*. We call such a graph a *tensegrity graph*. Moreover, we call the pair of a graph G and a configuration p a *tensegrity*, and we denote it as $G(p)$. For any $m \in \mathbb{N}_0$, let $q = (q_1, q_2, \ldots, q_k)$ be another configuration, where $q_1, q_2, \ldots, q_k \in \mathbb{E}^m$. The tensegrity

$G(q)$ is said to satisfy the *tensegrity constraints* of $G(p)$ if the following two conditions hold:

(c) If $\{i, j\}$ is a cable, then the inequality $\|q_i - q_j\| \leq \|p_i - p_j\|$ holds:
(s) If $\{i, j\}$ is a strut, then the inequality $\|q_i - q_j\| \geq \|p_i - p_j\|$ is true.

The tensegrity $G(p)$ is said to be *unyielding* if for any other configuration q in $\mathbb{E}^m$ such that $G(q)$ satisfies the tensegrity constraints of $G(p)$, then $\|q_i - q_j\| = \|p_i - p_j\|$ for every cable or strut $\{i, j\}$ of G. We note that in the definition of unyielding, the target configuration q can lie in any Euclidean space of arbitrary dimension.

One of the simplest examples of unyielding tensegrities $G(p)$ has its configuration consisting of three different collinear points, $p = (p_1, p_2, p_3)$, where p_2 lies between p_1 and p_3, and its tensegrity graph G is defined to have cables $\{1, 2\}$, $\{2, 3\}$ and a strut $\{1, 3\}$. Moreover, here is a simple but more interesting tensegrity: Let $p = (p_1, p_2, p_3, p_4)$ be a configuration consisting of the vertices, in cyclic order, of a rectangle in $\mathbb{E}^2$. We define the external edges $\{1, 2\}$, $\{2, 3\}$, $\{3, 4\}$, and $\{4, 1\}$ to be cables and the diagonals $\{1, 3\}$, $\{2, 4\}$ to be struts for the tensegrity graph G. Then, by Beckman and Quarles [3], $G(p)$ is a unyielding tensegrity.

Assume that $f : \mathbb{E}^n \to \mathbb{E}^m$ is a mapping, where n is an integer greater than 1. Let us define

$$C_f = \big\{c \in \mathbb{R} : \|f(x) - f(y)\| \leq c \text{ for all } x, y \in \mathbb{E}^n \text{ with } \|x - y\| = c\big\},$$
$$S_f = \big\{s \in \mathbb{R} : \|f(x) - f(y)\| \geq s \text{ for all } x, y \in \mathbb{E}^n \text{ with } \|x - y\| = s\big\}.$$

We call C_f the *cable lengths* for f and S_f the *strut lengths* for f.Indeed, C_f is considered as the set of all contractive distances by f and S_f as the set of all extensive distances by f. We note that both C_f and S_f include all positive real numbers if and only if f is an isometry.

Suppose a set C of positive real numbers is a subset of C_f and another set S is a subset of S_f. If some other real number c belongs to C_f, then c is said to be an *implied cable length* for f. Analogously, if some other real number s belongs to S_f, then s is called an *implied strut length* for f. For example, if c_1 and c_2 are cable lengths for f, then $c_1 + c_2$ is an implied cable length for f.

K. Bezdek and R. Connelly [8] proved the following lemmas.

Lemma 8.14 *Given an integer $n > 1$, let $f : \mathbb{E}^n \to \mathbb{E}^m$ be a mapping, and let $G(p)$ be an unyielding tensegrity in $\mathbb{E}^n$ with cables of lengths $c_1, c_2, \ldots$ and with struts of lengths $s_1, s_2, \ldots$*

(i) *If $c_2, c_3, \ldots \in C_f$ and $s_1, s_2, \ldots \in S_f$, then the length c_1 is an implied strut length for f.*
(ii) *If $c_1, c_2, \ldots \in C_f$ and $s_2, s_3, \ldots \in S_f$, then the length s_1 is an implied cable length for f.*

Proof We first consider the case (i), where $c_2, c_3, \ldots \in C_f$ and $s_1, s_2, \ldots \in S_f$. We claim that c_1 is a strut length for f. On the contrary, assume that c_1 is not a strut length for f. Suppose there are points $u, v \in \mathbb{E}^n$ with $\|u-v\| = c_1$ but $\|f(u)-f(v)\| < c_1$. Construct a configuration $\hat{p}$ in $\mathbb{E}^n$ that is congruent to p, where $\hat{p}_i = u$ and $\hat{p}_j = v$ for some $i, j \in \mathbb{N}$. Then, $\|\hat{p}_i - \hat{p}_j\| = c_1$. We define a configuration $q = (f(\hat{p}_1), f(\hat{p}_2), \ldots, f(\hat{p}_k))$, where k is the number of vertices of G. Then, we have

$$\|p_i - p_j\| = \|\hat{p}_i - \hat{p}_j\| = \|u - v\| = c_1 > \|f(u) - f(v)\| = \|q_i - q_j\|.$$

Since the cable lengths of G are in C_f, if $\{i', j'\}$ is the other cable in G, then we get

$$\|p_{i'} - p_{j'}\| = \|\hat{p}_{i'} - \hat{p}_{j'}\| \geq \|f(\hat{p}_{i'}) - f(\hat{p}_{j'})\| = \|q_{i'} - q_{j'}\|.$$

Since the strut lengths of G are in S_f, if $\{i', j'\}$ is a strut in G, then we obtain

$$\|p_{i'} - p_{j'}\| = \|\hat{p}_{i'} - \hat{p}_{j'}\| \leq \|f(\hat{p}_{i'}) - f(\hat{p}_{j'})\| = \|q_{i'} - q_{j'}\|.$$

Hence, $G(q)$ satisfies the tensegrity constraints of $G(p)$, but the strict inequality contradicts the unyielding property of $G(p)$. Therefore, our assumption is wrong, and we conclude that $\|f(u)-f(v)\| \geq c_1$ for each $u, v \in \mathbb{E}^n$ with $\|u-v\| = c_1$, which completes the proof of (i).

The other case (ii) can be proven in a similar way. □

Lemma 8.15 *Let $f : \mathbb{E}^n \to \mathbb{E}^m$ be a mapping, where n is an integer greater than 1. If c_2 and s_1 are positive real numbers with $c_2 \leq s_1 \leq 2c_2$, $c_2 \in C_f$, and $s_1 \in S_f$, then $c_1 = \frac{s_1^2 - c_2^2}{c_2}$ is an implied strut length in S_f.*

Proof Let $G(p)$ be the tensegrity shown in the following figure.

Consider the isosceles triangle defined by three edges of lengths c_2, c_2, and s_1, where s_1 is the base length. (This is possible because $s_1 \leq 2c_2$.) Place the two isosceles triangles as shown in Fig. 8.1. In this case, these four points form a trapezoid with diagonals of length s_1 if and only if $c_2 \leq s_1$. The equality cases correspond to degenerate cases where the configuration becomes collinear (when $s_1 = 2c_2$) and when the two vertices at the bottom of Fig. 8.1 coincide (when $c_2 = s_1$). It is a simple calculation to see, under these assumptions, that $c_1 = \frac{s_1^2 - c_2^2}{c_2}$.

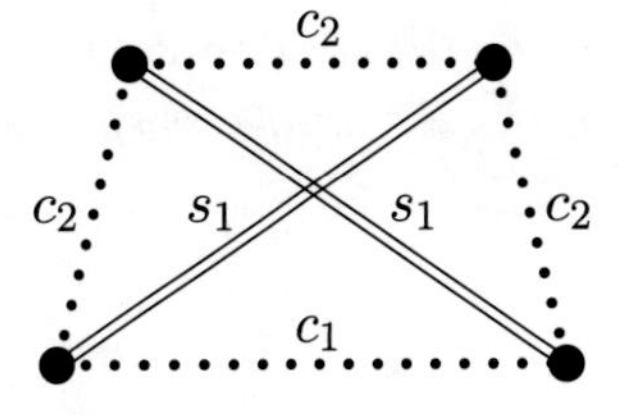

Fig. 8.1 In this figure, c_2 is a contractive distance by f and s_1 is an extensive distance by f

Based on the results in [14], the tensegrity graph in Fig. 8.1 is unyielding. Since c_1 is the length of a cable shown in Fig. 8.1, by Lemma 8.14, it is an implied strut length for every mapping $f : \mathbb{E}^n \to \mathbb{E}^m$. □

Let ϕ be the golden ratio, i.e., $\phi = \frac{\sqrt{5}+1}{2}$.

Lemma 8.16 *Let* $g : [1, 2] \to [0, 3]$ *be the mapping defined by* $g(t) = t^2 - 1$. *For every* $t_0 \in (\phi, 2]$, *a finite iteration of* t_0 *through* g *lies in the interval* $[2, 3]$.

Proof It is easy to show that g is strictly monotonically increasing and continuous and that $g^{-1}(t) = \sqrt{t+1}$. It is also obvious that the only fixed point for g (and g^{-1}) is the solution of the quadratic equation $t^2 - 1 = t$ $(t \geq 1)$. Thus, the unique fixed point for g is the golden ratio ϕ. We note that ϕ is a repulsive fixed point for g but an attractive fixed point for g^{-1}. Indeed, if any t in the domain of g satisfies $t > \phi$, then $g(t) > t$. Since $g(t) > t$ for $t > \phi$, and $g(\phi) = \phi$, and since g is monotonically increasing, it follows that $\phi < g^{-1}(t) < t$ for $t > \phi$.

We now set $\alpha_0 = 3$ and inductively $\alpha_i = g^{-1}(\alpha_{i-1})$ for all $i \in \mathbb{N}$. Thus, we have $\alpha_1 = 2$, $\alpha_2 = \sqrt{3}$, $\alpha_3 = \sqrt{\sqrt{3}+1}, \ldots$ By the argument above, $\{\alpha_i\}$ is a well-defined decreasing sequence, where each $\alpha_i > \phi$. So this sequence must converge to a fixed point of g. Therefore, since ϕ is the unique fixed point in the domain of g, we conclude that $\lim_{i\to\infty} \alpha_i = \phi$.

If we consider images of intervals under g^{-1}, we then have

$$[2, 3] \to \left[\sqrt{3}, 2\right] \to \left[\sqrt{\sqrt{3}+1}, \sqrt{3}\right] \to \cdots$$

In other words, $g^{-1}([\alpha_i, \alpha_{i-1}]) = [g^{-1}(\alpha_i), g^{-1}(\alpha_{i-1})] = [\alpha_{i+1}, \alpha_i]$ for $i \in \mathbb{N}$. But this implies

$$\bigcup_{i=1}^{\infty} [\alpha_i, \alpha_{i-1}] = (\phi, 3],$$

which completes the proof. □

At a seminar held in 1986, F. Rádo et al. [53] presented the following excellent result.

Theorem 8.17 (Rádo, Andreason and Válcan) *Let* n *be an integer greater than* 1, *and let* c, s *be positive real numbers that satisfy* $0 < \frac{c}{s} \leq \frac{1}{\sqrt{3}}$. *Assume that a mapping* f : $\mathbb{E}^n \to \mathbb{E}^m$ *satisfies the following conditions:*

(i) If $\|x - y\| = c$ *for* $x, y \in \mathbb{E}^n$, *then* $\|f(x) - f(y)\| \leq c$;
(ii) If $\|x - y\| = s$ *for* $x, y \in \mathbb{E}^n$, *then* $\|f(x) - f(y)\| \geq s$.

Then f *is an isometry.*

Using both conditions given in Theorem 8.17, we define

$$F_k(c, s) = \{f : \mathbb{E}^n \to \mathbb{E}^m : f \text{ satisfies the conditions } (i) \text{ and } (ii) \text{ for all } n \geq k\},$$

where k is a fixed positive integer. Then the set we want to identify is:

$$X_k = \left\{r \in \mathbb{R} : \text{If } \frac{c}{s} = r, \text{ then each } f \in F_k(c, s) \text{ is an isometry}\right\}.$$

So Theorem 8.17 simply says that $(0, \frac{1}{\sqrt{3}}] \subset X_2$. It is easy to check that

$$X_2 \subset X_3 \subset X_4 \subset \cdots.$$

We are now ready to prove the main theorem of this section.

Theorem 8.18 (Bezdek and Connelly) $(0, \frac{1}{\phi}) \subset X_2$.

Proof Assume that $0 < c < s$, $0 < \frac{c}{s} < \frac{1}{\phi}$, $f : \mathbb{E}^n \to \mathbb{E}^m$ is a mapping for $n > 1$, and that $f \in F_2(c, s)$. In other words, c is a cable length for f and s is a strut length for f. Rescaling if necessary, we can assume $c = 1$ without loss of generality.

If $\frac{c}{s} \leq \frac{1}{2} < \frac{1}{\sqrt{3}}$, then $\frac{1}{s} \leq \frac{1}{2}$ and $s \geq 2$. Therefore, f is an isometry by Theorem 8.17. If $\frac{1}{2} \leq \frac{c}{s} = \frac{1}{s} < \frac{1}{\phi}$, then $\phi < s \leq 2$. It follows from Lemma 8.16 that a finite iteration of s under g lies in the interval $[2, 3]$, where we set $g(t) = t^2 - 1$. We set $s_0 = s$ and $s_i = g(s_{i-1})$ for $i \in \mathbb{N}$. From Lemma 8.15, it follows that if $s_{i-1} \in S_f$, then $s_i = s_{i-1}^2 - 1 = g(s_{i-1}) \in S_f$, too. Hence, for some positive integer k, s_k is an implied strut in S_f, and $s_k \in [2, 3]$. Therefore, using Theorem 8.17, we conclude that f is an isometry. These imply that $s \in X_2$, which completes the proof. □

The following corollary is a direct consequence of Theorems 2.7 and 8.18.

Corollary 8.19 (Bezdek and Connelly) *Let* m *and* n *be integers with* $1 < n \leq m$. *If a mapping* $f : \mathbb{E}^n \to \mathbb{E}^m$ *preserves two distances* ρ *and* σ, *where* $0 < \frac{\rho}{\sigma} < \frac{\sqrt{5}-1}{2}$, *then* f *is an affine isometry.*

8.3 Aleksandrov-Benz Problem with Restricted Domains

In 1987, W. Benz and H. Berens [7] presented a sufficient condition for a mapping with both contractive and extensive distances to be an affine isometry (see Theorem 3.5):

Let X be a real normed space with $\dim X > 1$ *and let Y be a real normed space that is strictly convex. Suppose* $f : X \to Y$ *is a mapping and* $N > 1$ *is a fixed integer. If a distance* $\rho > 0$ *is contractive and* $N\rho$ *is extensive by f, then f is an affine isometry.*

In this section, we present sufficient conditions for the Benz-Berens theorem to hold even in a restricted domain.

Let H be a real Hilbert space with $\dim H > 2$ for which there exists a unit vector $w \in H$ and a subspace H_s of H with $H = H_s \oplus \mathrm{span} w$ and $H_s \perp \mathrm{span} w$, where $\mathrm{span} w$ denotes the subspace spanned by w. We now define the half space H_θ by

$$H_\theta = \{x + \lambda w : x \in H_s,\ \lambda > \theta\} \tag{8.1}$$

for any real number θ. Assume that Y is a real normed space which is strictly convex.

Throughout this section, let a real number $\rho > 0$ and an integer $N > 1$ be fixed. Furthermore, assume that a mapping $f : H_\theta \to Y$ satisfies both the following conditions:

$(P1)$ ρ is contractive by f;
$(P2)$ $N\rho$ is extensive by f.

Following the steps presented in the paper [4], S.-M. Jung and Th. M. Rassias [39] proved in the following two lemmas that if a mapping $f : H_\theta \to Y$ satisfies both conditions $(P1)$ and $(P2)$, then f preserves the distances ρ and 2ρ.

Lemma 8.20 *Assume that a mapping* $f : H_\theta \to Y$*, together with a real number* $\rho > 0$ *and an integer* $N > 1$*, satisfies both conditions* $(P1)$ *and* $(P2)$*, where* H_θ *is given by* (8.1) *and Y is a real normed space that is strictly convex. Then* $\|f(x) - f(y)\| = \rho$ *holds for all* $x, y \in H_\theta$ *with* $\|x - y\| = \rho$.

Proof Assume that x and y are points of H_θ with $\|x - y\| = \rho$ and $x - y \in \overline{H}_0$, where we set

$$\overline{H}_0 = \{x + \lambda w : x \in H_s, \lambda \geq 0\}.$$

If we define $p_n = y + n(x - y)$ for any $n \in \{0, 1, \ldots, N\}$, then $p_n \in H_\theta$, $\|p_N - y\| = N\rho$ and $\|p_n - p_{n-1}\| = \rho$ for all $n \in \{1, 2, \ldots, N\}$. Since f satisfies both conditions $(P1)$ and $(P2)$, we have

$$N\rho \le \|f(p_N) - f(y)\| \le \sum_{n=1}^{N} \|f(p_n) - f(p_{n-1})\| \le N\rho.$$

Hence, we conclude that $\|f(x) - f(y)\| = \|f(p_1) - f(p_0)\| = \rho$.

Now assume that x and y of H_θ satisfy $\|x - y\| = \rho$ but $x - y \notin \overline{H}_0$. Then we have $y - x \in \overline{H}_0$. In this case, if we define $p_n = x + n(y - x)$, we get the same result using a similar method as before. □

Lemma 8.21 *Assume that a mapping $f : H_\theta \to Y$, together with a real number $\rho > 0$ and an integer $N > 1$, satisfies both conditions $(P1)$ and $(P2)$, where H_θ is given by* (8.1) *and Y is a real normed space that is strictly convex. Then $\|f(x) - f(y)\| = 2\rho$ holds for all $x, y \in H_\theta$ with $\|x - y\| = 2\rho$.*

Proof Assume that x and y are points of H_θ with $\|x - y\| = 2\rho$ and $x - y \in \overline{H}_0$, where we may refer to the proof of Lemma 8.20 for the definition of $\overline{H}_0$. Let us define

$$p_n = y + \frac{n}{2}(x - y)$$

for all $n \in \{0, 1, \ldots, N\}$. It then follows that $p_n \in H_\theta$, $\|p_N - y\| = N\rho$, and $\|p_n - p_{n-1}\| = \rho$ for all $n \in \{1, 2, \ldots, N\}$. Now, we make use of $(P1)$ and $(P2)$ to get

$$N\rho \le \|f(p_N) - f(y)\| \le \sum_{n=1}^{N} \|f(p_n) - f(p_{n-1})\| \le N\rho,$$

i.e.,

$$\|f(p_N) - f(y)\| = \sum_{n=1}^{N} \|f(p_n) - f(p_{n-1})\|. \tag{8.2}$$

If we assume

$$\|f(p_2) - f(p_0)\| < \|f(p_2) - f(p_1)\| + \|f(p_1) - f(p_0)\|,$$

then, in view of (8.2), it should be $N > 2$, and further

$$\begin{aligned} \|f(p_N) - f(y)\| &\le \sum_{n=3}^{N} \|f(p_n) - f(p_{n-1})\| + \|f(p_2) - f(p_0)\| \\ &< \sum_{n=1}^{N} \|f(p_n) - f(p_{n-1})\|, \end{aligned}$$

which is contrary to (8.2). Therefore, we conclude by Lemma 8.20 that

$$\begin{aligned}\|f(x) - f(y)\| &= \|f(p_2) - f(p_0)\| \\ &= \|f(p_2) - f(p_1)\| + \|f(p_1) - f(p_0)\| \\ &= 2\rho.\end{aligned}$$

For the case of $x - y \notin \overline{H}_0$, we define $p_n = x + \frac{n}{2}(y - x)$ and follow the same process as before to prove our assertion. □

Due to the strict convexity of Y, the following lemma is an immediate consequence of Lemma 2.6. Therefore, the proof is omitted.

Lemma 8.22 *Assume that Y is a real normed space that is strictly convex. For all $a, b, c \in Y$ and $\alpha > 0$, $\|b - a\| = \alpha = \|c - b\|$ and $\|c - a\| = 2\alpha$ imply $c = 2b - a$.*

We will use mathematical induction to prove the following theorem, which is essential to handle the case where x and y have the same H_s-components. From now on, we denote by x_s, y_s, and z_s the H_s-component of x, y and z, respectively, unless specified.

Lemma 8.23 *Assume that a mapping $f : H_\theta \to Y$, together with a real number $\rho > 0$ and an integer $N > 1$, satisfies both conditions $(P1)$ and $(P2)$, where H_θ is given by (8.1) and Y is a real normed space that is strictly convex. For any given $n \in \mathbb{N}$, let $x = x_s + \lambda w$ and $y = y_s + \mu w$ be any points of H_θ with $x_s = y_s$ and*

$$\lambda, \mu > \theta + \left(\frac{1}{2^2} + \frac{1}{2^3} + \cdots + \frac{1}{2^{n+1}}\right)\rho.$$

Then, $\|x - y\| = \frac{1}{2^n}\rho$ implies $\|f(x) - f(y)\| = \frac{1}{2^n}\rho$.

Proof Assume that $x = x_s + \lambda w$ and $y = y_s + \mu w$ are points of H_θ such that

$$x_s = y_s, \quad \lambda, \mu > \theta + \frac{1}{4}\rho, \quad \|x - y\| = |\lambda - \mu| = \frac{1}{2}\rho.$$

Choose a $z = z_s + \frac{1}{2}(\lambda + \mu)w \in H_\theta$ with $\|x - z\| = \|y - z\| = \rho$. Furthermore, select x' and y' on the rays $\overline{zx}$ and $\overline{zy}$, respectively, such that $\|x' - z\| = \|y' - z\| = 2\rho$. Then, we have $\|x' - y'\| = \rho$.

If we set $x' = x'_s + \lambda' w$ and $y' = y'_s + \mu' w$, then

$$\lambda' = \lambda + \frac{1}{2}(\lambda - \mu) > \left(\theta + \frac{1}{4}\rho\right) + \frac{1}{2}\left(-\frac{1}{2}\rho\right) = \theta$$

and

$$\mu' = \mu + \frac{1}{2}(\mu - \lambda) > \left(\theta + \frac{1}{4}\rho\right) + \frac{1}{2}\left(-\frac{1}{2}\rho\right) = \theta.$$

Thus, it follows that both x' and y' are points in H_θ.

According to Lemmas 8.20 and 8.21, we get

$$\begin{aligned}
\|f(x) - f(z)\| &= \|f(y) - f(z)\| = \|f(x') - f(y')\| = \rho, \\
\|f(x') - f(x)\| &= \|f(y') - f(y)\| = \rho, \\
\|f(x') - f(z)\| &= \|f(y') - f(z)\| = 2\rho.
\end{aligned}$$

By Lemma 8.22, $f(x)$ is the midpoint of $f(x')$ and $f(z)$, and the same is true for $f(y)$. Hence, the triangles $f(x)f(z)f(y)$ and $f(x')f(z)f(y')$ are similar. Hence, we conclude that $\|f(x) - f(y)\| = \frac{1}{2}\rho$.

We now assume that our assertion holds for some $n \in \mathbb{N}$ and moreover assume that $x = x_s + \lambda w$ and $y = y_s + \mu w$ satisfy

$$x_s = y_s, \quad \lambda, \mu > \theta + \left(\frac{1}{2^2} + \frac{1}{2^3} + \cdots + \frac{1}{2^{n+2}}\right)\rho, \quad \|x - y\| = \frac{1}{2^{n+1}}\rho.$$

Choose a $z = z_s + \frac{1}{2}(\lambda + \mu)w$ with $\|x - z\| = \|y - z\| = \rho$. Moreover, select x' and y' on the rays $\overline{zx}$ and $\overline{zy}$, respectively, such that $\|x' - z\| = \|y' - z\| = 2\rho$. Then we obtain $\|x' - y'\| = \frac{1}{2^n}\rho$. Similarly as in the first part, it follows that both the x' and y' lie in H_θ.

By Lemmas 8.20 and 8.21, we get

$$\begin{aligned}
\|f(x) - f(z)\| &= \|f(y) - f(z)\| = \rho, \\
\|f(x') - f(x)\| &= \|f(y') - f(y)\| = \rho, \\
\|f(x') - f(z)\| &= \|f(y') - f(z)\| = 2\rho.
\end{aligned}$$

By Lemma 8.22, $f(x)$ is a midpoint of $f(x')$ and $f(z)$, and likewise for $f(y)$. Furthermore, we know that $x' = x_s' + \lambda' w$ and $y' = y_s' + \mu' w$ satisfy

$$x_s' = y_s', \quad \lambda', \mu' > \theta + \left(\frac{1}{2^2} + \frac{1}{2^3} + \cdots + \frac{1}{2^{n+1}}\right)\rho, \quad \|x' - y'\| = \frac{1}{2^n}\rho.$$

By the assumption of induction, it follows that $\|f(x') - f(y')\| = \frac{1}{2^n}\rho$.

Since the triangles $f(x)f(z)f(y)$ and $f(x')f(z)f(y')$ are similar, we conclude that $\|f(x) - f(y)\| = \frac{1}{2^{n+1}}\rho$. □

In the following lemma, we prove that if x and y are separated from each other by a specific distance, then some equidistant points on the line through x and y are mapped by f onto some equidistant points of the line through $f(x)$ and $f(y)$.

Lemma 8.24 *Assume that a mapping $f : H_\theta \to Y$, together with a real number $\rho > 0$ and an integer $N > 1$, satisfies both conditions $(P1)$ and $(P2)$, where H_θ is given by* (8.1) *and Y is a real normed space that is strictly convex.*

(i) *If x and y are any points of H_θ with $\|x - y\| = \rho$, then $f(x + m(y - x)) = f(x) + m(f(y) - f(x))$ holds for all $m \in \mathbb{N}_0$ with $x + m(y - x) \in H_\theta$.*

(ii) *Let x, y be points of $H_{\theta+\rho/2}$ with $x_s = y_s$ and $\|x - y\| = \frac{1}{2^n}\rho$ for some $n \in \mathbb{N}$. If $x + m(y - x) \in H_{\theta+\rho/2}$ for some $m \in \mathbb{N}$, then $f(x + m(y - x)) = f(x) + m(f(y) - f(x))$.*

Proof

(i) Assume that x and y are points of H_θ with $\|x - y\| = \rho$. We use induction to show that $f(x+m(y-x)) = f(x)+m(f(y)-f(x))$ holds for all $m \in \mathbb{N}_0$ with $x+m(y-x) \in H_\theta$. If $m = 0$ or 1, there is nothing to prove. We now assume that our assertion is true for $m \in \{0, 1, \ldots, k\}$, where $k > 0$ is some integer. We set $p_i = x + i(y - x)$ for $i \in \mathbb{N}$ and assume that $p_{k+1} \in H_\theta$. Then, we get

$$\|p_k - p_{k-1}\| = \rho = \|p_{k+1} - p_k\| \quad \text{and} \quad \|p_{k+1} - p_{k-1}\| = 2\rho.$$

According to Lemmas 8.20 and 8.21, we have

$$\|f(p_k) - f(p_{k-1})\| = \rho = \|f(p_{k+1}) - f(p_k)\|,$$
$$\|f(p_{k+1}) - f(p_{k-1})\| = 2\rho.$$

Hence, it follows from Lemma 8.22 that

$$f(p_{k+1}) = 2f(p_k) - f(p_{k-1}) = f(x) + (k + 1)\big(f(y) - f(x)\big),$$

as we desired.

(ii) Let $x = x_s + \lambda w$ and $y = y_s + \mu w$ be any points of $H_{\theta+\rho/2}$. Assume that $x_s = y_s$ and $\|x - y\| = \frac{1}{2^n}\rho$ for some $n \in \mathbb{N}$. We also use induction to prove our assertion. For $m = 1$, there is nothing to prove. We assume that our assertion holds for $m \in \{1, 2, \ldots, k\}$, where $k > 0$ is some integer. We now set $p_i = x + i(y - x)$ for $i \in \mathbb{N}$ and let $p_{k+1} \in H_{\theta+\rho/2}$. Then, we have

$$\|p_k - p_{k-1}\| = \frac{1}{2^n}\rho = \|p_{k+1} - p_k\| \quad \text{and} \quad \|p_{k+1} - p_{k-1}\| = \frac{1}{2^{n-1}}\rho.$$

Since the H_s-component of p_i is equal to x_s and $p_i \in H_{\theta+\rho/2}$ for each $i \in \{1, 2, \ldots, k+1\}$, we can make use of Lemma 8.23 to check that

$$\|f(p_k) - f(p_{k-1})\| = \frac{1}{2^n}\rho = \|f(p_{k+1}) - f(p_k)\|,$$

$$\|f(p_{k+1}) - f(p_{k-1})\| = \frac{1}{2^{n-1}}\rho.$$

Hence, it follows from Lemma 8.22 that

$$f(p_{k+1}) = 2f(p_k) - f(p_{k-1}) = f(x) + (k+1)\big(f(y) - f(x)\big),$$

which completes the proof of (ii). □

Lemma 8.25 *Assume that a mapping $f : H_\theta \to Y$, together with a real number $\rho > 0$ and an integer $N > 1$, satisfies both conditions $(P1)$ and $(P2)$, where H_θ is given by (8.1) and Y is a real normed space that is strictly convex. Let n be a fixed positive integer. If $x, y \in H_\theta$ satisfy $\|x - y\| = n\rho$, then $\|f(x) - f(y)\| = n\rho$.*

Proof Assume that x and y are points of H_θ and are separated from each other by a distance $n\rho$. Choose a point z on the line segment between x and y such that $x = y + n(z - y)$. Then, we have $\|z - y\| = \rho$. From Lemma 8.24 (i), it follows that $f(x) = f(y) + n(f(z) - f(y))$. Hence, by Lemma 8.20, we have

$$\|f(x) - f(y)\| = n\|f(z) - f(y)\| = n\rho,$$

which completes the proof. □

Using Lemmas 8.23, 8.24, and 8.25, we can prove the following lemma which is indispensable for the proof of Theorem 8.28 below.

Lemma 8.26 *Assume that a mapping $f : H_\theta \to Y$, together with a real number $\rho > 0$ and an integer $N > 1$, satisfies both conditions $(P1)$ and $(P2)$, where H_θ is given by (8.1) and Y is a real normed space that is strictly convex. Let $x = x_s + \lambda w$ and $y = y_s + \mu w$ be any points of H_θ. Assume that $m, n \in \mathbb{N}$ are given.*

(i) If $x_s \neq y_s$ and $\|x - y\| = \frac{n}{m}\rho$, then $\|f(x) - f(y)\| = \frac{n}{m}\rho$.
(ii) If $x, y \in H_{\theta+\rho/2}$, $x_s = y_s$, and $\|x - y\| = \frac{m}{2^n}\rho$, then $\|f(x) - f(y)\| = \frac{m}{2^n}\rho$.

Proof

(i) Assume that x and y are points of H_θ with $\|x - y\| = \frac{n}{m}\rho$ which are represented by $x = x_s + \lambda w$ and $y = y_s + \mu w$, where $x_s \neq y_s$, $\lambda \geq \mu > \theta$, and where $m > 1$ and n are positive integers.

Set $z = z_s + \mu w$, and examine whether there exists a $z_s \in H_s$ which is a solution of the following parametric equations

$$\|z - x\|^2 = \|z_s - x_s\|^2 + (\mu - \lambda)^2 = k^2\rho^2,$$
$$\|z - y\|^2 = \|z_s - y_s\|^2 = k^2\rho^2,$$
$$\|x - y\|^2 = \|x_s - y_s\|^2 + (\mu - \lambda)^2 = \left(\frac{n}{m}\rho\right)^2,$$

where k is a parameter whose value is integral. It follows from these equations that

$$\begin{aligned} \|z_s - x_s\| &= \sqrt{k^2\rho^2 - (\mu - \lambda)^2}, \\ \|z_s - y_s\| &= k\rho, \\ \|x_s - y_s\| &= \sqrt{\left(\frac{n}{m}\rho\right)^2 - (\mu - \lambda)^2}. \end{aligned} \tag{8.3}$$

The sphere in H_s with radius $\sqrt{k^2\rho^2 - (\mu - \lambda)^2}$ and center at x_s is expressed by the first equation of (8.3). We use the notation S_1 for this sphere. The second of (8.3) is an equation for the sphere S_2 in H_s with radius $k\rho$ and center at y_s. If k is so large that the inequality

$$k\rho \leq \sqrt{k^2\rho^2 - (\mu - \lambda)^2} + \sqrt{\left(\frac{n}{m}\rho\right)^2 - (\mu - \lambda)^2}$$

holds, then $S_1 \cap S_2 \neq \emptyset$. Hence, we can select a z_s from $S_1 \cap S_2$, i.e., the parametric equations (8.3) are solvable in z_s. With such a z_s, $z = z_s + \mu w$ is separated from x resp. from y by a same distance $k\rho$.

If we choose $x', y' \in H_\theta$ on the ray $\overline{zx}$ resp. $\overline{zy}$ such that $\|x' - z\| = \|y' - z\| = km\rho$, then we have $\|x' - y'\| = n\rho$. By Lemma 8.25, we get

$$\|f(x) - f(z)\| = \|f(y) - f(z)\| = k\rho,$$
$$\|f(x') - f(z)\| = \|f(y') - f(z)\| = km\rho,$$
$$\|f(x') - f(y')\| = n\rho.$$

Furthermore, by a slight modification of Lemma 8.24 (i), we can conclude that $f(x)$ lies on the line segment between $f(z)$ and $f(x')$ and also that $f(y)$ lies on the line segment between $f(z)$ and $f(y')$.

Hence, the triangles $f(x)f(z)f(y)$ and $f(x')f(z)f(y')$ are similar. Therefore, we obtain $\|f(x) - f(y)\| = \frac{n}{m}\rho$.

(ii) Assume that $x = x_s + \lambda w$ and $y = y_s + \mu w$ are points of $H_{\theta+\rho/2}$ with $x_s = y_s$ and $\|x - y\| = \frac{m}{2^n}\rho$. Choose a z on the line segment between x and y with $\|z - y\| = \frac{1}{2^n}\rho$. Then, by Lemma 8.23, $\|f(z) - f(y)\| = \frac{1}{2^n}\rho$. Further, in view of Lemma 8.24 (ii), we get

$$f(x) = f\big(y + m(z - y)\big) = f(y) + m\big(f(z) - f(y)\big),$$

i.e.,

$$\|f(x) - f(y)\| = m\|f(z) - f(y)\| = \frac{m}{2^n}\rho,$$

which completes the proof. □

Lemma 8.27 *Let H_θ be defined by* (8.1). *Assume that α and β are real numbers with $2\beta \geq \alpha > 0$. Then, for all $x, y \in H_\theta$ with $\|x - y\| = \alpha$, there exists a $z \in H_\theta$ satisfying $\|z - x\| = \beta = \|z - y\|$. In particular, if $x_s \neq y_s$, then $z_s \notin \{x_s, y_s\}$.*

Proof Assume that $x = x_s + \lambda w$ and $y = y_s + \mu w$ are points of H_θ with $\|x - y\| = \alpha$, where $\lambda, \mu > \theta$. We find a solution $z = z_s + \delta w \in H_\theta$ of the following equations:

$$\begin{aligned} \|z - x\|^2 &= \|z_s - x_s\|^2 + (\delta - \lambda)^2 = \beta^2, \\ \|z - y\|^2 &= \|z_s - y_s\|^2 + (\delta - \mu)^2 = \beta^2, \\ \|x - y\|^2 &= \|x_s - y_s\|^2 + (\lambda - \mu)^2 = \alpha^2. \end{aligned} \tag{8.4}$$

Put $\delta = \frac{1}{2}(\lambda + \mu)$ $(> \theta)$. It then follows from (8.4) that

$$\begin{aligned} \|z_s - x_s\|^2 &= \beta^2 - \frac{1}{4}(\mu - \lambda)^2, \\ \|z_s - y_s\|^2 &= \beta^2 - \frac{1}{4}(\mu - \lambda)^2, \\ \|x_s - y_s\|^2 &= \alpha^2 - (\mu - \lambda)^2. \end{aligned}$$

Since $\dim H_s > 1$ and

$$\begin{aligned} \|z_s - x_s\| + \|z_s - y_s\| &= 2\|z_s - x_s\| \\ &= \sqrt{(2\beta)^2 - (\mu - \lambda)^2} \\ &\geq \sqrt{\alpha^2 - (\mu - \lambda)^2} \\ &= \|x_s - y_s\| \end{aligned}$$

(where $\|x_s - y_s\| > 0$ for $x_s \neq y_s$, and hence $z_s \neq x_s$ and $z_s \neq y_s$), there exists at least one $z_s \in H_s$ which is a solution of the above equations. With such a z_s, $z = z_s + \frac{1}{2}(\lambda + \mu)w \in H_\theta$ satisfies our requirement. Therefore, the proof is complete. □

So far, we have proved all the preliminary lemmas to the main theorem of this section. In the following theorem, we generalize the theorem of Benz and Berens:

Theorem 8.28 (Jung and Rassias) *Let H be a real Hilbert space with* $\dim H > 2$. *Assume that H_θ is given by* (8.1) *and Y is a real normed space that is strictly convex. Let $\rho > 0$ be a real number and let $N > 1$ be an integer. If ρ is contractive and $N\rho$ is extensive by a mapping $f : H_\theta \to Y$, then $f|_{H_{\theta+\rho/2}}$ is an isometry. In particular, it holds that $\|f(x) - f(y)\| = \|x - y\|$ for all points $x, y \in H_\theta$ with $x_s \neq y_s$.*

Proof We note that there exists a unit vector $w \in H$ and a subspace $H_s = (\mathrm{span} w)^\perp$ of H with $H = H_s \oplus \mathrm{span} w$ and $H_s \perp \mathrm{span} w$, where $(\mathrm{span} w)^\perp$ denotes the orthogonal complement of $\mathrm{span} w$.

Assume that x and y are distinct points of $H_{\theta+\rho/2}$. For those x and y, choose the sequences, $\{k_i\}$, $\{m_i\}$, and $\{n_i\}$, of nonnegative integers with the following three properties:

(K) $\frac{k_i}{2^{n_i}}\rho \leq \|x - y\| < \frac{k_i+1}{2^{n_i}}\rho$ for all sufficiently large integers i;
(M) $\frac{m_i-1}{2^{n_i}}\rho < \|x - y\| \leq \frac{m_i}{2^{n_i}}\rho$ for all sufficiently large integers i;
(N) $\{n_i\}$ increases strictly to infinity.

Since $H_{\theta+\rho/2}$ is open, we can select a z_i on the line segment $\overline{xy}$ and a $w_i \in H_{\theta+\rho/2}$ such that

$$\|x - z_i\| = \frac{k_i}{2^{n_i}}\rho \quad \text{and} \quad \|z_i - w_i\| = \|w_i - y\| = \frac{1}{2^{n_i}}\rho$$

for any sufficiently large integer i. It then follows from Lemma 8.26 (i) and (ii) that

$$\|f(x) - f(z_i)\| = \frac{k_i}{2^{n_i}}\rho \quad \text{and} \quad \|f(z_i) - f(w_i)\| = \|f(w_i) - f(y)\| = \frac{1}{2^{n_i}}\rho$$

for any sufficiently large integer i. Thus, it follows from (K) that

$$\begin{aligned}\|f(x) - f(y)\| &\leq \|f(x) - f(z_i)\| + \|f(z_i) - f(w_i)\| + \|f(w_i) - f(y)\| \\ &\leq \|x - y\| + \frac{1}{2^{n_i-1}}\rho\end{aligned}$$

for any sufficiently large integer i, i.e., we get $\|f(x) - f(y)\| \leq \|x - y\|$.

On the other hand, since $H_{\theta+\rho/2}$ is an open subset of the real Hilbert space H, we can choose a $v_i \in H_{\theta+\rho/2}$ such that

$$\|x - v_i\| = \frac{m_i}{2^{n_i}}\rho \quad \text{and} \quad \|y - v_i\| = \frac{1}{2^{n_i}}\rho$$

for all sufficiently large integers i. From Lemma 8.26 (i) and (ii), we get

$$\|f(x) - f(v_i)\| = \frac{m_i}{2^{n_i}}\rho \quad \text{and} \quad \|f(y) - f(v_i)\| = \frac{1}{2^{n_i}}\rho.$$

Hence, it follows from (M) that

$$\|f(x) - f(y)\| \geq \|f(x) - f(v_i)\| - \|f(y) - f(v_i)\| \geq \|x - y\| - \frac{1}{2^{n_i}}\rho$$

for all sufficiently large integers i, i.e., we get $\|f(x) - f(y)\| \geq \|x - y\|$, which completes the proof of the first part.

For the second part of this theorem, assume that $x, y \in H_\theta$ satisfy $x_s \neq y_s$ and $r_1\rho < \|x - y\| < r_2\rho$, where $r_1, r_2 > 0$ are given rational numbers. We claim that $r_1\rho \leq \|f(x) - f(y)\| \leq r_2\rho$: According to Lemma 8.27, there exists a $z \in H_\theta$ with $\|z - x\| = \frac{r_2}{2}\rho = \|z - y\|$, $x_s \neq z_s$, and $y_s \neq z_s$. Due to Lemma 8.26 (i), we get

$$\|f(z) - f(x)\| = \frac{r_2}{2}\rho = \|f(z) - f(y)\|.$$

Hence, we have

$$\|f(x) - f(y)\| \leq \|f(x) - f(z)\| + \|f(z) - f(y)\| = r_2\rho.$$

On the other hand, we assume that there existed $x, y \in H_\theta$ with the properties:

$$x_s \neq y_s, \quad r_1\rho < \|x - y\| < r_2\rho, \quad \|f(x) - f(y)\| < r_1\rho. \tag{8.5}$$

Then, we obtain

$$r_2\rho - \|x - y\| < r_2\rho - r_1\rho < r_2\rho - \|f(x) - f(y)\|.$$

Define $z = x + \lambda(y - x)$ for the case $y - x \in \overline{H_0}$ with $\lambda = \frac{r_2}{\|x-y\|}\rho > 1$. (Otherwise, i.e., if $y - x \notin \overline{H_0}$, we replace the definition of z by $y + \lambda(x - y)$ and repeat the following process similarly.) It then follows that $x_s \neq z_s$, $y_s \neq z_s$, and $\|z - x\| = r_2\rho$. Furthermore, (8.5) implies that $\|z - y\| = (\lambda - 1)\|x - y\| < (r_2 - r_1)\rho$. Due to Lemma 8.26 (i), we have $\|f(z) - f(x)\| = r_2\rho$ and by considering the argument in the last paragraph, we see that $\|f(z) - f(y)\| \leq (r_2 - r_1)\rho$. Subsequently, we have

$$\begin{aligned} r_2\rho &= \|f(z) - f(x)\| \\ &\le \|f(z) - f(y)\| + \|f(y) - f(x)\| \\ &< (r_2 - r_1)\rho + r_1\rho \\ &= r_2\rho, \end{aligned}$$

which is a contradiction. Therefore, it should be $r_1\rho \le \|f(x) - f(y)\| \le r_2\rho$.

Since the set of all rational numbers is dense in $\mathbb{R}$, we conclude that the second assertion is true. □

Let H be a real Hilbert space with $\dim H > 1$. For a fixed integer $N > 1$ and a constant $\rho > 0$, we define a sequence $\{d_i\}$ by

$$d_1 = N\rho \quad \text{and} \quad d_i = N^{3-i}\rho$$

for all $i \in \{2, 3, \ldots\}$.

Let $\{H_i\}$ be a sequence of open convex subsets of H with

$$H_0 \supset H_1 \supset \cdots \supset H_i \supset H_{i+1} \supset \cdots \quad \text{and} \quad d(H_{i+1}, \partial H_i) \ge d_{i+1}$$

for all $i \in \mathbb{N}_0$, where we set

$$d(H_{i+1}, \partial H_i) = \inf\left\{\|x - y\| : x \in H_{i+1},\ y \in \partial H_i\right\}$$

and ∂H_i denotes the boundary of H_i. (If one of H_{i+1} and ∂H_i is unbounded, we will set $d(H_{i+1}, \partial H_i) = \infty$.)

Furthermore, we assume

$$H_\infty := \left(\bigcap_{i=0}^{\infty} H_i\right)^{\circ} \ne \emptyset.$$

We know that the intersection of any family of convex subsets of a topological vector space is convex. Moreover, the interior of any convex subset of a topological vector space is a convex set. Thus, H_∞ is an open convex subset of H.

S.-M. Jung proved, in his paper [29], the following theorem, which in some sense generalizes Theorem 8.28 (compare with the result in [40]).

Theorem 8.29 (Jung) *Let H and H' be real Hilbert spaces with* $\dim H > 1$. *Assume that H_0 is a nonempty open convex subset of H. If a mapping $f : H_0 \to H'$ satisfies both the conditions* (P1) *and* (P2), *then $f|_{H_\infty}$ is an isometry.*

8.4 Beckman-Quarles Theorem with Restriced Domains

In this section, let $\mathbb{E}^n$ be the n-dimensional Euclidean space, where $n > 2$ is a fixed integer. Then there exists a unit vector $w \in \mathbb{E}^n$ and a subspace E_s of $\mathbb{E}^n$ such that $\mathbb{E}^n = E_s \oplus \text{span}w$ and E_s is orthogonal to span$w$, where span$w$ is the subspace of $\mathbb{E}^n$ which is spanned by w. (We simply take $E_s = (\text{span}w)^\perp$.)

We now define

$$r_0 = \theta, \quad r_1 = \theta + \rho, \quad r_2 = \theta + \rho + \rho_1, \quad r_3 = \theta + \left(1 + \frac{1}{n}\right)\rho + \rho_1,$$

where θ is a real number, ρ is a positive real number, and $\rho_1 = \sqrt{2(1 + \frac{1}{n})}\rho$. Using these r_k's, we further define

$$E_k = \{x + \lambda w : x \in E_s,\ \lambda > r_k\}$$

for $k \in \{0, 1, 2, 3\}$. We note that

$$E_3 \subset E_2 \subset E_1 \subset E_0 \subset \mathbb{E}^n.$$

Let E be a nonempty subset of $\mathbb{E}^n$. We call a set of n different points of E a β-set in E if the distance between any two points is $\beta > 0$. If there are two distinct points of $\mathbb{E}^n$ that have the distance α from each point of a β-set P, the two points are called the α-associated points of P.

Due to Lemmas 2.21 and 2.22, the following two lemmas are obvious.

Lemma 8.30 *Let E be a nonempty subset of $\mathbb{E}^n$, where $n > 2$. Assume that α and β are positive real numbers with*

$$\gamma(\alpha, \beta) = 4\alpha^2 - 2\beta^2\left(1 - \frac{1}{n}\right) > 0$$

and that P is a β-set in E. The α-associated points of P are uniquely determined and the distance between them is $\sqrt{\gamma(\alpha, \beta)}$.

Lemma 8.31 *Assume that α and β are positive real numbers with $\gamma(\alpha, \beta) > 0$. If x and y are points of $\mathbb{E}^n$ with $\|x - y\| = \sqrt{\gamma(\alpha, \beta)}$, then there exists a β-set P in $\mathbb{E}^n$ such that x and y are the α-associated points of P.*

Lemma 8.32 *If a mapping $f : E_0 \to \mathbb{E}^n$ $(n > 2)$ preserves a distance $\rho > 0$, then the distance $\rho_1 = \sqrt{\gamma(\rho, \rho)}$ is preserved by $f|_{E_1}$.*

Proof Assume that x, y are points of E_1 with $\|x - y\| = \rho_1$. According to Lemma 8.31 and the definition of E_k, there exists a ρ-set P in E_0 such that x and y are the ρ-associated points of P. Since f preserves ρ, $P' = f(P)$ is also a ρ-set in $\mathbb{E}^n$.

Due to Lemma 8.30, there are exactly two distinct ρ-associated points x' and y' of P', and they satisfy $\|x' - y'\| = \sqrt{\gamma(\rho, \rho)} = \rho_1$. Since there exist only two ρ-associated points of P', we have $\{f(x), f(y)\} \subset \{x', y'\}$, i.e., $\|f(x) - f(y)\| = 0$ or ρ_1.

Assume that $f(x) = f(y)$. Choose a $z \in E_0$ with $\|x - z\| = \rho_1$ and $\|y - z\| = \rho$. According to Lemma 8.31, there exists a ρ-set Q in E_0 such that x and z are the ρ-associated points of Q (Because $x \in E_1$ and $\|x - q\| = \rho$ for each $q \in Q$, Q is a subset of E_0). Similarly, $Q' = f(Q)$ is a ρ-set in $\mathbb{E}^n$.

Due to Lemma 8.30, there exist exactly two distinct ρ-associated points x'' and z'' of Q' which satisfy $\|x'' - z''\| = \sqrt{\gamma(\rho, \rho)} = \rho_1$. Hence, $\{f(x), f(z)\} \subset \{x'', z''\}$, i.e., $\|f(x) - f(z)\| = 0$ or ρ_1, i.e., $\|f(y) - f(z)\| = 0$ or ρ_1, because we assumed $f(x) = f(y)$.

On the other hand, we obtain $\rho = \|y - z\| = \|f(y) - f(z)\| = 0$ or ρ_1, which is a contradiction. Altogether, we conclude that $\|f(x) - f(y)\| = \rho_1$. □

Lemma 8.33 *If a mapping $f : E_0 \to \mathbb{E}^n$ $(n > 2)$ preserves a distance $\rho > 0$, then the distance $\rho_2 = \sqrt{\gamma(\rho_1, \rho_1)} = \frac{2(n+1)}{n}\rho$ is preserved by $f|_{E_2}$.*

Proof Assume that x and y are points of E_2 that satisfy $\|x - y\| = \rho_2$. According to Lemma 8.31, there exists a ρ_1-set P in E_1 such that x and y are the ρ_1-associated points of P. Since $f|_{E_1}$ preserves ρ_1 by Lemma 8.32, $P' = f(P)$ is also a ρ_1-set in $\mathbb{E}^n$.

According to Lemma 8.30, there exist only two distinct ρ_1-associated points x' and y' of P' whose distance is $\|x' - y'\| = \rho_2$. Hence, it follows that $\{f(x), f(y)\} \subset \{x', y'\}$, i.e., $\|f(x) - f(y)\| = 0$ or ρ_2.

Assume that $f(x) = f(y)$. Choose a $z \in E_1$ with $\|x - z\| = \rho_2$ and $\|y - z\| = \rho_1$. (Because of $y \in E_2$ and $\|y - z\| = \rho_1$, we conclude that $z \in E_1$.) In view of Lemma 8.31, there exists a ρ_1-set Q in E_1 such that x and z are the ρ_1-associated points of Q. (Because $x \in E_2$ and $\|x - q\| = \rho_1$ for all $q \in Q$, Q is a subset of E_1.) Hence, $Q' = f(Q)$ is a ρ_1-set in $\mathbb{E}^n$ (see Lemma 8.32).

By Lemma 8.30, there exist exactly two distinct ρ_1-associated points x'' and z'' of Q' and $\|x'' - z''\| = \rho_2$. Therefore, we have $\|f(x) - f(z)\| = 0$ or ρ_2, i.e., $\|f(y) - f(z)\| = 0$ or ρ_2, because we assumed $f(x) = f(y)$.

Since $y, z \in E_1$, by Lemma 8.32, we have $\rho_1 = \|y - z\| = \|f(y) - f(z)\| = 0$ or ρ_2, a contradiction. Altogether, we conclude that $\|f(x) - f(y)\| = \rho_2$. □

Lemma 8.34 *If a mapping $f : E_0 \to \mathbb{E}^n$ $(n > 2)$ preserves a distance $\rho > 0$, then the distance $\rho_3 = \sqrt{\gamma(\rho, \rho_1)} = \frac{2}{n}\rho$ is contractive by $f|_{E_2}$.*

Proof Assume that x and y are points of E_2 with $\|x - y\| = \rho_3$. By Lemma 8.31, there exists a ρ_1-set P in E_1 such that x and y are the ρ-associated points of P. (We note that

$x \in E_2$ and $\|x - p\| = \rho$ for all $p \in P$. Hence, P is a subset of E_1.) By Lemma 8.32, $P' = f(P)$ is also a ρ_1-set in $\mathbb{E}^n$.

According to Lemma 8.30, there exist only two distinct ρ-associated points x' and y' of P' with $\|x' - y'\| = \rho_3$. Hence, it follows that $\|f(x) - f(y)\| = 0$ or ρ_3. Consequently, we have $\|f(x) - f(y)\| \leq \rho_3$. □

We are ready to generalize the theorem of Beckman and Quarles by proving that if a mapping, from a half space E_0 of $\mathbb{E}^n$ into $\mathbb{E}^n$, preserves a distance $\rho > 0$, then the restriction of f to a half space E_3 is an isometry.

S.-M. Jung and Th. M. Rassias [39] proved the following theorem (see also [35]).

Theorem 8.35 (Jung and Rassias) *If a mapping $f : E_0 \to \mathbb{E}^n$ $(n > 2)$ preserves a distance $\rho > 0$, then the restriction $f|_{E_3}$ is an isometry. In particular, if any points x and y of E_2 satisfy $x_s \neq y_s$, where x_s and y_s are the E_s-components of x and y, then it holds that $\|f(x) - f(y)\| = \|x - y\|$.*

Proof According to Lemmas 8.33 and 8.34, the distance $\frac{2}{n}\rho$ is contractive, and the distance $\frac{2(n+1)}{n}\rho$ is extensive (preserved) by $f|_{E_2}$. Hence, by Theorem 8.28, the restriction $f|_{E_3}$ is an isometry. In view of the second part of Theorem 8.28, the second part of this theorem is obviously true. □

From now on, we assume that j and n are fixed integers with $0 \leq j \leq n$ and $n > 1$. Let $w_1, w_2, \ldots, w_j$ be some orthonormal vectors in $\mathbb{E}^n$, and let E_s be a subspace of $\mathbb{E}^n$ such that

$$\mathbb{E}^n = E_s \oplus (\mathrm{span} w_1) \oplus (\mathrm{span} w_2) \oplus \cdots \oplus (\mathrm{span} w_j).$$

We set $E_s = \mathbb{E}^n$ for $j = 0$.

Let us define a sequence $\{s_k\}$ of positive real numbers by

$$s_0 = 0, \quad s_1 = \rho, \quad s_2 = \left(1 + \sqrt{\frac{2(n+1)}{n}}\right)\rho,$$

$$s_k = \left(n + 2 + \sqrt{\frac{2(n+1)}{n}} + \sum_{i=4}^{k} \frac{1}{(n+1)^{i-5}}\right)\rho$$

for all $k \in \{3, 4, \ldots\}$, where $\rho > 0$ is a fixed real number. We denote by s_∞ the limit point of the sequence $\{s_k\}$, i.e.,

$$s_\infty = \lim_{k\to\infty} s_k = \left(n + 2 + \sqrt{\frac{2(n+1)}{n}} + \frac{(n+1)^2}{n}\right)\rho.$$

By C_k, we denote an open convex subset of $\mathbb{E}^n$ defined by

$$C_k = \{x + \lambda_1 w_1 + \cdots + \lambda_j w_j : x \in E_s,\ c_i + s_k < \lambda_i < d_i - s_k \\ \text{for } i \in \{1, 2, \ldots, j\}\}$$

for any $k \in \mathbb{N}_0 \cup \{\infty\}$, where $c_i, d_i \in \mathbb{R} \cup \{\pm\infty\}$ are constants with $d_i - c_i > 2s_\infty$ for every $i \in \{1, 2, \ldots, j\}$. We then note that

$$C_\infty \subset \cdots \subset C_{k+1} \subset C_k \subset \cdots \subset C_0 \subset \mathbb{E}^n.$$

S.-M. Jung [30] proved the following theorem.

Theorem 8.36 (Jung) *If a mapping $f : C_0 \to \mathbb{E}^n$ $(n > 1)$ preserves a distance $\rho > 0$, then the restriction $f|_{C_\infty}$ is an isometry.*

8.5 Beckman-Quarles Theorem with Geometric Figures

F. S. Beckman and D. A. Quarles [3] solved the Aleksandrov problem for finite-dimensional Euclidean spaces $\mathbb{E}^n$, where n is an integer greater than 1:

If a mapping $f : \mathbb{E}^n \to \mathbb{E}^n$ preserves a distance $\rho > 0$, then f is an affine isometry.

It seems interesting to investigate whether the "distance $\rho > 0$" in the Beckman-Quarles theorem can be replaced by some properties characterized by "geometric figures."

In this section, the triangles (quadrilaterals, pentagons, hexagons, and circles) denote the peripheries of the geometric figures. A side of a triangle (a quadrilateral, a pentagon, or a hexagon) without its endpoints (vertices) is called an *open side*.

Mappings Preserving Regular Triangles

For an integer $n > 1$ and a constant $r > 0$, we use T_r^n to denote the set of all regular triangles in $\mathbb{E}^n$ whose side length is r. We note that if two regular triangles $T_1, T_2 \in T_r^n$ intersect each other in an infinite number of points distributed on two open sides of T_1, then T_2 is either coincident with T_1 or a shift of T_1 along a side of T_1 including infinitely many intersection points. In the latter case, T_2 must intersect one of the open sides of T_1 in exactly one point.

The following theorem was proved by S.-M. Jung [26].

Theorem 8.37 (Jung) *Let m, n be integers greater than 1 and let a, b be positive constants. If $f : \mathbb{E}^m \to \mathbb{E}^n$ is a one-to-one mapping under which the image of each regular triangle in T_a^m belongs to T_b^n, then the equality $\|f(x) - f(y)\| = b$ holds for all $x, y \in \mathbb{E}^m$ with $\|x - y\| = a$.*

Proof Let x and y be any points of $\mathbb{E}^m$ with $\|x-y\| = a$. We choose another point $z \in \mathbb{E}^m$ such that the three points comprise the vertices of a triangle $T_1 \in T_a^m$. Furthermore, we choose a point $w \in \mathbb{E}^m \setminus \{x\}$ coplanar with x, y, z such that y, z, w are the vertices of a triangle $T_2 \in T_a^m$.

Let x', y', z' be the vertices of the image of the triangle T_1 under f. Assume that the image of the side $\overline{yz}$ is spread out on the open sides $\overline{x'y'}$ and $\overline{y'z'}$. If each of the open sides includes more than one image point of $\overline{yz}$, then the image of the triangle T_2 would coincide with the image of T_1, which would be a contradiction to the injectivity of f. Thus, we assume without loss of generality that $\overline{x'y'}$ contains infinitely many image points of $\overline{yz}$ and the open side $\overline{y'z'}$ contains only one image point of $\overline{yz}$, say u', and let u be the unique point of $\overline{yz}$ satisfying $u' = f(u)$, where u is allowed to be y or z.

We choose another regular triangle $T_3 \in T_a^m$, which is different from T_1 and T_2, with the following properties:

(i) T_3 is coplanar with T_1 and T_2;
(ii) T_3 contains infinitely many points of $\overline{yz}$. In particular, T_3 contains u.

Then, the image of T_3 under f would coincide with the image of T_1 or T_2, which would also be a contradiction to the injectivity of f.

By applying the same argument, we see that f maps each side of T_1 into one of the sides of its image triangle. Because of the hypothesis, $f(T_1) \in T_b^n$, f maps each side of T_1 onto a corresponding side of its image triangle.

We note that x is the unique intersection point of the sides $\overline{zx}$ and $\overline{xy}$ of T_1, and likewise for y and z. Thus, $f(x), f(y), f(z)$ should be the vertices of the image of T_1 under f, which belongs to T_b^n. Altogether, we conclude that $f(x)$ is separated from $f(y)$ by a distance b, which completes the proof. □

In the following corollary, we prove that if a one-to-one mapping $f : \mathbb{E}^n \to \mathbb{E}^n$ maps the periphery of every regular triangle of side length $a > 0$ onto the periphery of a regular triangle of side length $b > 0$, then there exists an affine isometry $I : \mathbb{E}^n \to \mathbb{E}^n$ such that $f(x) = \frac{b}{a} I(x)$.

Corollary 8.38 *Let n be an integer greater than 1, and let a, b be some real positive constants. If $f : \mathbb{E}^n \to \mathbb{E}^n$ is a one-to-one mapping under which the image of each regular triangle from T_a^n belongs to T_b^n, then there exists an affine isometry $I : \mathbb{E}^n \to \mathbb{E}^n$ such that*

$$f(x) = \frac{b}{a} I(x)$$

for all $x \in \mathbb{E}^n$.

Proof If we define a mapping $I : \mathbb{E}^n \to \mathbb{E}^n$ by $I(x) = \frac{a}{b} f(x)$ for all $x \in \mathbb{E}^n$, then it follows from Theorem 8.37 that I satisfies $\|I(x) - I(y)\| = a$ for all $x, y \in \mathbb{E}^n$ with $\|x - y\| = a$. According to the Beckman-Quarles theorem, the mapping I is an affine isometry, and hence the assertion of this corollary is proved. □

Mappings Preserving Regular Quadrilaterals

By $\mathcal{Q}_r^n$, we denote the set of all regular quadrilaterals (squares) in $\mathbb{E}^n$ whose side length is r, where n is an integer greater than 1 and r is a positive constant. We remember that quadrilateral means the periphery of the quadrilateral.

We remark that when two regular quadrilaterals $Q_1, Q_2 \in \mathcal{Q}_r^n$ intersect in an infinite number of points distributed on two adjacent open sides of Q_1, they actually coincide. The proof of the following theorem was presented in [26].

Theorem 8.39 (Jung) *Let m, n be integers greater than 1, and let a, b be positive constants. If $f : \mathbb{E}^m \to \mathbb{E}^n$ be a one-to-one mapping under which the image of each regular quadrilateral in $\mathcal{Q}_a^m$ belongs to $\mathcal{Q}_b^n$, then the equality $\|f(x) - f(y)\| = b$ holds for all $x, y \in \mathbb{E}^m$ with $\|x - y\| = a$.*

Proof Let p_1 and p_2 be arbitrary points in $\mathbb{E}^m$ that satisfy $\|p_1 - p_2\| = a$. We then choose $p_3, p_4 \in \mathbb{E}^m$ such that p_1, p_2, p_3, and p_4 comprise the vertices of a regular quadrilateral $Q_0 \in \mathcal{Q}_a^m$. Moreover, for every $i \in \{1, 2, 3, 4\}$, we choose a regular quadrilateral Q_i that is coplanar with Q_0 and has the common side $\overline{p_i p_{i+1}}$ with Q_0, where we set $p_5 = p_1$, as shown in Fig. 8.2.

Let Q_i' be the image of Q_i under f for every $i \in \{0, 1, 2, 3, 4\}$. According to the hypotheses, the Q_i''s are regular quadrilaterals of $\mathcal{Q}_b^n$ with the following properties:

(i) $Q_0' \cap Q_i'$ is an infinite set for $i \in \{1, 2, 3, 4\}$;
(ii) $Q_1' \cap Q_2'$, $Q_2' \cap Q_3'$, $Q_3' \cap Q_4'$ and $Q_4' \cap Q_1'$ are one-point sets;
(iii) Q_0' is a proper subset of $Q_1' \cup Q_2' \cup Q_3' \cup Q_4'$;
(iv) Q_i' and Q_{i+2}' are disjoint for $i \in \{1, 2\}$.

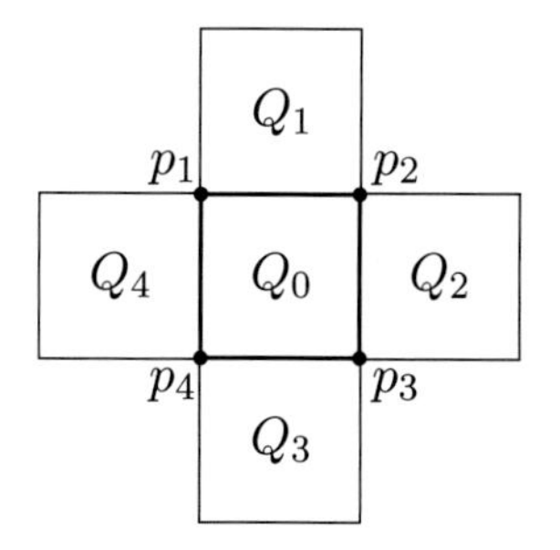

Fig. 8.2 Each of the regular quadrilaterals Q_1, Q_2, Q_3, Q_4 is coplanar with Q_0 and intersects Q_0 in a corresponding whole side of Q_0

The vertices of Q_0' are denoted by p_1', p_2', p_3', p_4'. The side $\overline{p_1p_2}$ is the only common side of Q_0 and Q_1. If the image of $\overline{p_1p_2}$ were spread out on two adjacent open sides of Q_0', then Q_1' would coincide with Q_0' as we mentioned in the sentence immediately before this theorem, which would contradict the injectivity of f.

Assume now that the image of the side $\overline{p_1p_2}$ is spread out on both parallel open sides $\overline{p_1'p_2'}$ and $\overline{p_3'p_4'}$ of Q_0'. Since the injectivity of f causes Q_1' not to coincide with Q_0', the Q_1' has to be a shift of Q_0' along the sides $\overline{p_1'p_2'}$ and $\overline{p_3'p_4'}$. In view of (i) and (iii), and by considering the above argument, one of Q_2', Q_3', Q_4' should meet Q_0' in two parallel (opposite) sides of Q_0', since each side of Q_0' contains a nondegenerate line segment that is a part of $Q_0' \setminus Q_1'$. In this case, however, the $Q_0', Q_1', \ldots, Q_4'$ cannot satisfy all the conditions (i), (ii), (iii), and (iv).

Therefore, the image of $\overline{p_1p_2}$ under f should be included in one side of Q_0'. In general, every side of Q_0 should be mapped into (and hence onto) a corresponding side of Q_0'.

Since p_1, p_2, p_3, p_4 are unique intersection points of the sides $\overline{p_4p_1}$ and $\overline{p_1p_2}$, $\overline{p_1p_2}$ and $\overline{p_2p_3}$, $\overline{p_2p_3}$ and $\overline{p_3p_4}$, $\overline{p_3p_4}$ and $\overline{p_4p_1}$, respectively, the points $f(p_1)$, $f(p_2)$, $f(p_3)$, $f(p_4)$ comprise the vertices of Q_0' in the same (or opposite) cyclic order as Q_0. Therefore, $f(p_1)$ is separated from $f(p_2)$ by the distance b. □

In the following corollary, we prove that if a one-to-one mapping $f : \mathbb{E}^n \to \mathbb{E}^n$ maps the periphery of every regular quadrilateral of side length $a > 0$ onto the periphery of a regular quadrilateral of side length $b > 0$, then there exists an affine isometry $I : \mathbb{E}^n \to \mathbb{E}^n$ such that $f(x) = \frac{b}{a} I(x)$.

Corollary 8.40 *Let n be an integer greater than 1, and let $a, b > 0$ be some constants. If $f : \mathbb{E}^n \to \mathbb{E}^n$ is a one-to-one mapping under which the image of each regular quadrilateral from Q_a^n belongs to Q_b^n, then there is an affine isometry $I : \mathbb{E}^n \to \mathbb{E}^n$ such that*

$$f(x) = \frac{b}{a} I(x)$$

for all $x \in \mathbb{E}^n$.

Proof We define a mapping $I : \mathbb{E}^n \to \mathbb{E}^n$ by $I(x) = \frac{a}{b} f(x)$ for all $x \in \mathbb{E}^n$. It then follows from Theorem 8.39 that I satisfies $\|I(x) - I(y)\| = a$ for all $x, y \in \mathbb{E}^n$ with $\|x - y\| = a$. By the Beckman-Quarles theorem, I is an affine isometry. □

Mappings Preserving Regular Pentagons

We now use P_r^n to denote the set of all regular pentagons in $\mathbb{E}^n$ whose side length is r, where $n > 1$ is an integer and $r > 0$ is a real number.

Assume that $P_1, P_2 \in P_r^n$. The following statement is obviously true:

(P) *If* P_1 *intersects* P_2 *in an entire side, except for a finite number of points, and in a point on another open side, then* P_1 *coincides with* P_2.

Let L be a line segment in $\mathbb{E}^n$ with length r. If we denote by $|P_1 \cap L|$, the number of elements of $P_1 \cap L$, then

$$|P_1 \cap L| \in \{0, 1, 2, \infty\},$$

where $|P_1 \cap L| = \infty$ holds if and only if L lies upon a substantial part of a side of P_1.

Assume that P_1 intersects P_2 in three distinct points which lie on a side of P_2. If we select this side of P_2 and call it L, then it follows from our assumption that $|P_1 \cap L| > 2$, which is equivalent to $|P_1 \cap L| = \infty$. This argument can be restated as follows:

(Q) *If* P_1 *intersects* P_2 *in three distinct points on a side of* P_2, *then* P_1 *intersects* P_2 *in a line segment containing those three points.*

If P_1 intersects P_2 in four different points, three points of which lie on one side of P_2 and the remaining point is not collinear with the three points, then either P_2 coincides with P_1 or P_2 is a shift of P_1 along the side containing the three points, as we see in Fig. 8.3:

According to the above conclusion, we can easily see that:

(R) P_1 *coincides with* P_2 *if* P_1 *intersects two different sides of* P_2 *in an infinite number of points each.*

Now assume that P_1 intersects P_2 in an infinite number of points, including two vertices of P_2 that are not adjacent to each other. Choose a side of P_2, so that P_1 intersects that side in at least three different points. Then P_1 additionally intersects P_2 in a vertex of P_2, and this vertex is not collinear with those three points. In this case, P_1 coincides with P_2, as shown in Fig. 8.4:

The above argument may be summarized as follows:

(S) *If* P_1 *intersects* P_2 *in an infinite number of points, including two vertices of* P_2 *that are not adjacent, then* P_1 *coincides with* P_2.

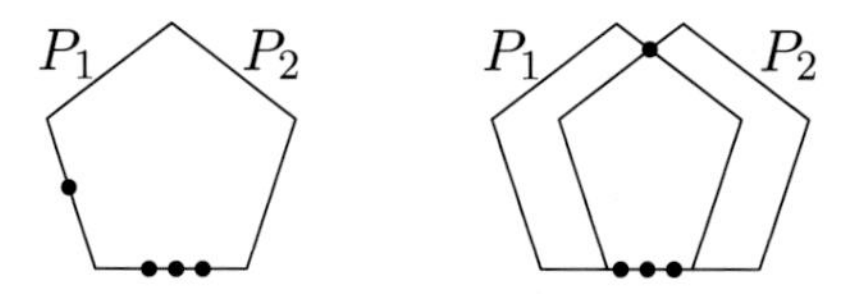

Fig. 8.3 Three of the intersection points lie on one side of P_2, and the remaining point lies either on the adjacent or the far side of P_2

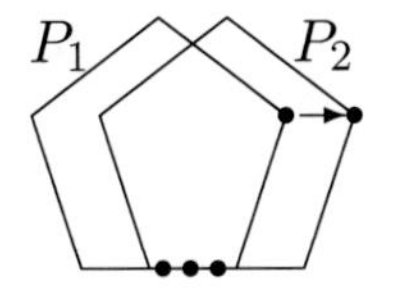

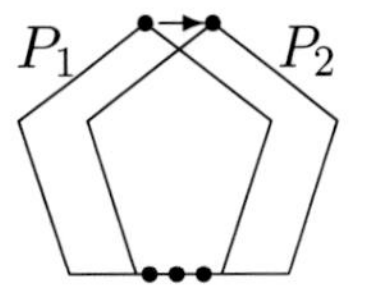

Fig. 8.4 If P_1 intersects P_2 in three points on a side of P_2 and in a vertex that is not collinear with these three points, then they are identical

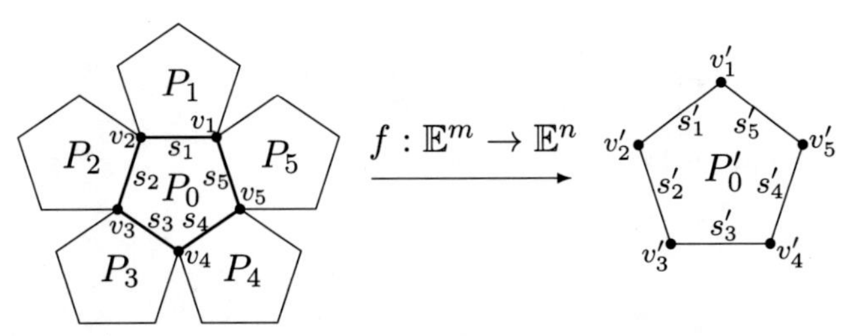

Fig. 8.5 The image P'_i of a regular pentagon P_i under f is also a regular pentagon for all $i \in \{0, 1, \ldots, 5\}$

From now on, for given integers $m, n > 1$ and real constants $a, b > 0$, let $f : \mathbb{E}^m \to \mathbb{E}^n$ be a one-to-one mapping under which the image of every regular pentagon in P_a^m belongs to P_b^n.

We now consider the following figure:

On the left side of Fig. 8.5, six regular pentagons are labeled $P_0, P_1, \ldots, P_5$, all of which belong to P_a^m. The pentagon P_0 consists of five sides, denoted $s_1, s_2, \ldots, s_5$. Each side s_i is a line segment containing only one point v_i of its two endpoints (vertices).

Due to our hypothesis, the image of any regular pentagon P_i under f is also a regular pentagon, and we use the notation P'_i for the image of P_i, i.e., $P'_i = f(P_i)$ for each $i \in \{0, 1, \ldots, 5\}$. Furthermore, we use the notations s'_i and v'_i for the sides and vertices of P'_0, respectively. Each side s'_i is a line segment containing only one point v'_i of its two endpoints. (At this point, we are not sure that all the v'_i are images of the v_j, and likewise for the s'_i.)

The following lemma is obvious.

Lemma 8.41 *It holds that*

$$P'_0 = \bigcup_{k=1}^{5} s'_k \subset \bigcup_{k=1}^{5} P'_k$$

under the above assumptions.

Since f is a one-to-one mapping, considering (R), we can easily verify the following lemma. We therefore omit the proof.

Lemma 8.42 *For every* $i \in \{1, 2, \ldots, 5\}$*, there exists a unique* $j \in \{1, 2, \ldots, 5\}$ *such that* $s'_j \cap P'_i$ *is an infinite set, but* $(P'_0 \setminus s'_j) \cap P'_i$ *is a finite set.*

Conversely, we can prove the following lemma.

Lemma 8.43 *For every* $i \in \{1, 2, \ldots, 5\}$*, there is a* $j \in \{1, 2, \ldots, 5\}$ *such that* $s'_i \cap P'_j$ *is an infinite set, but* $s'_i \cap \bigcup_{k \neq j} P'_k$ *is a finite set.*

Proof For every fixed $i \in \{1, 2, \ldots, 5\}$, it follows from Lemma 8.41 that

$$s'_i \subset \bigcup_{k=1}^{5} P'_k \quad \text{or} \quad s'_i = \bigcup_{k=1}^{5} (s'_i \cap P'_k).$$

Since s'_i contains infinitely many points, there must be a $j \in \{1, 2, \ldots, 5\}$ such that $s'_i \cap P'_j$ is an infinite set.

We now prove the second claim. Without loss of generality, we assume that both $s'_1 \cap P'_1$ and $s'_1 \cap P'_2$ were infinite sets. It would then follow from (R) and Lemmas 8.41 and 8.42 that

$$\begin{aligned}
\bigcup_{k=1}^{5} s'_k = P'_0 &= \bigcup_{k=1}^{5} (P'_0 \cap P'_k) \\
&= (s'_1 \cap P'_1) \cup (s'_1 \cap P'_2) \cup \bigcup_{k=3}^{5} (P'_0 \cap P'_k) \\
&\quad \cup \left\{\text{finite number of points on } P'_0 \setminus s'_1\right\} \\
&= \left(s'_1 \cap (P'_1 \cup P'_2)\right) \cup (s'_{i_1} \cap P'_3) \cup (s'_{i_2} \cap P'_4) \cup (s'_{i_3} \cap P'_5) \\
&\quad \cup \left\{\text{finite number of points on } P'_0\right\} \\
&\neq \bigcup_{k=1}^{5} s'_k,
\end{aligned}$$

which would be a contradiction, where $i_1, i_2, i_3 \in \{1, 2, \ldots, 5\}$ are chosen such that $s'_{i_1} \cap P'_3$, $s'_{i_2} \cap P'_4$ and $s'_{i_3} \cap P'_5$ are infinite sets. □

According to (P), two congruent regular pentagons are identical if they have an entire side and a point on another open side in common.

In the following lemma, we prove that f maps every vertex v_i of P_0 to a vertex v'_j of $P'_0 = f(P_0)$.

Lemma 8.44 *For every $i \in \{1, 2, \ldots, 5\}$, there exists a $j \in \{1, 2, \ldots, 5\}$ with $f(v_i) = v_j'$.*

Proof Without loss of generality, we assume that $f(v_1) \in s_1' \setminus \{v_1'\}$. (We assume that $f(v_1)$ lies on the open side s_1' without endpoints.) According to Lemma 8.43, there exists a $j \in \{1, 2, \ldots, 5\}$ such that $s_5' \cap P_j'$ is an infinite set but $s_5' \cap \bigcup_{k \neq j} P_k'$ is a finite set. Without loss of generality, we may choose 5 for such j, i.e., we assume that $s_5' \cap P_5'$ is an infinite set (see Fig. 8.5). Since $P_0' \neq P_5'$, it follows from Lemma 8.42 that both the sets

$$\bigcup_{k=1}^{4} s_k' \cap P_5' \quad \text{and} \quad s_5' \cap \bigcup_{k=1}^{4} P_k'$$

are finite sets. Since $s_5' \cap P_5'$ is an infinite set and $s_5' \cap P_5' \subset P_0' \cap P_5'$, the property (Q) implies that P_5' intersects P_0' in a line segment L including $s_5' \cap P_5'$ such that $(s_5' \setminus L) \cap P_5' = \emptyset$.

Suppose $s' = s_5' \setminus L$ is an infinite set, i.e., a line segment of positive length. However, it follows that $s' = s_5' \setminus L \subset s_5' \setminus (s_5' \cap P_5') = s_5' \setminus P_5'$, and hence $s' \cap P_5' = \emptyset$. Thus, we have

$$s' = s' \cap P_0' \subset s' \cap \bigcup_{k=1}^{5} P_k' = s' \cap \bigcup_{k=1}^{4} P_k' \subset s_5' \cap \bigcup_{k=1}^{4} P_k',$$

and the last set is finite, while s' is assumed to be an infinite set. It clearly leads to a contradiction. Hence, P_5' intersects P_0' in a whole side s_5' except for finitely many points. Since $f(v_1) \in P_0' \cap P_5'$ (see Fig. 8.5), it follows from (P) that $P_0' = P_5'$, which obviously contradicts the fact that f is a one-to-one mapping. Therefore, we conclude that $f(v_1) = v_1'$. □

Using Lemmas 8.42 and 8.44 together with (S), we can now prove that the images of v_1 and v_2 are two vertices of P_0', which are adjacent to each other.

Lemma 8.45 *It holds that $\|f(v_1) - f(v_2)\| = b$.*

Proof It follows from Lemma 8.44 that $f(v_1), f(v_2) \in \{v_1', v_2', \ldots, v_5'\}$. Assume that $f(v_1)$ and $f(v_2)$ are two vertices of P_0' that are not adjacent to each other. Since $v_1, v_2 \in P_0 \cap P_1$, we have $f(v_1), f(v_2) \in P_0' \cap P_1'$. Since s_j' is a side of P_0', Lemma 8.42 implies that the congruent regular pentagons P_0' and P_1' intersect each other in infinitely many points including $f(v_1)$ and $f(v_2)$ which are vertices of P_0' not adjacent to each other. In view of (S), P_0' should coincide with P_1'. This would contradict the fact that f is a one-to-one mapping. Hence, $f(v_1)$ and $f(v_2)$ are two vertices of P_0' that are adjacent to each other. Therefore, we can conclude that $\|f(v_1) - f(v_2)\| = b$. (We recall that the side length of P_0' is b.) □

By referring to Lemmas 8.41 through 8.45, we can prove the following theorem presented in [28].

Theorem 8.46 (Jung) *Let n be an integer greater than 1, and let $a, b > 0$ be some real constants. If $f : \mathbb{E}^n \to \mathbb{E}^n$ is a one-to-one mapping under which the image of every regular pentagon of P_a^n belongs to P_b^n, then there is an affine isometry $I : \mathbb{E}^n \to \mathbb{E}^n$ such that*

$$f(x) = \frac{b}{a} I(x)$$

for all $x \in \mathbb{E}^n$.

Proof We define a mapping $I : \mathbb{E}^n \to \mathbb{E}^n$ by $I(x) = \frac{a}{b} f(x)$ for all $x \in \mathbb{E}^n$. For any two points $v_1, v_2 \in \mathbb{E}^n$ with $\|v_1 - v_2\| = a$, we can construct six regular pentagons $P_0, P_1, \ldots, P_5$ as indicated on the left side of Fig. 8.5. It then follows from Lemma 8.45 that

$$\|I(v_1) - I(v_2)\| = \frac{a}{b} \|f(v_1) - f(v_2)\| = a.$$

Hence, the mapping I preserves the distance a. According to the Beckman-Quarles theorem, I is an affine isometry. □

Mappings Preserving Regular Hexagons

Let H_r^n be the set of all regular hexagons in $\mathbb{E}^n$ whose side length is r, where n is an integer greater than 1 and r is a positive constant.

We note that if two different regular hexagons $H_1, H_2 \in H_r^n$ intersect each other in an infinite number of points distributed on two open sides of H_1, these two open sides lie opposite each other and H_2 is a shift of H_1 along these two open sides of H_1.

The following theorem was presented in [26].

Theorem 8.47 (Jung) *Let m, n be integers greater than 1, and let a, b be positive real constants. If $f : \mathbb{E}^m \to \mathbb{E}^n$ is a one-to-one mapping under which the image of every regular hexagon in H_a^m belongs to H_b^n, then the equality $\|f(x) - f(y)\| = b$ holds for all $x, y \in \mathbb{E}^m$ with $\|x - y\| = a$.*

Proof

(*a*) Let $p_1, p_2, \ldots, p_6 \in \mathbb{E}^m$ be vertices of a regular hexagon $P \in H_a^m$ with center at p_0. Furthermore, we assume that the point sets

$$\{p_0, p_3, q_3, q_4, q_5, p_5\}, \quad \{r_1, p_1, p_0, p_5, r_5, r_6\}, \quad \{s_1, s_2, s_3, p_3, p_0, p_1\}$$

are the vertices of regular hexagons $Q, R, S \in H_a^m$ with centers at p_4, p_6, and p_2, respectively (see Fig. 8.6).

Let P', Q', R', and S' be regular hexagons in H_b^n which are the images of P, Q, R, and S under f, respectively. Since p_0 is the only intersection point of Q, R, and S, and since the sides $\overline{p_0p_5}$, $\overline{p_0p_1}$, and $\overline{p_0p_3}$ are the intersections of Q and R, R, and S, S, and Q, respectively, the injectivity of f and the hypothesis imply that

(i) $Q' \cap R' \cap S'$ is a one-point set;
(ii) $Q' \cap R'$, $R' \cap S'$, $S' \cap Q'$ are infinite sets.

(b) Since f is one-to-one, there are only two ways for Q' to intersect S' in an infinite number of points:

- $Q' \cap S'$ is contained in one side of Q';
- $Q' \cap S'$ is contained in two opposite sides of Q'.

We note that two regular hexagons $Q', S' \in H_b^n$ that intersect in an infinite number of points are coincident in other cases. The above statements also apply to Q' and R' or R' and S'.

(c) We will first prove that the image hexagons Q', R' and S' are coplanar. For example, assume that Q' and R' are not coplanar. According to (ii) and (b), only one side of Q' overlaps with only one side of R'. (Otherwise, they would be coplanar.) Let the line segment $\overline{q'r'}$ be the nondegenerate intersection of Q' and R', where q' is a vertex of Q' and r' is a vertex of R' (see Fig. 8.7).

Also assume that S' has common points with Q' on at least two open sides of Q'. It then follows from (ii) and (b) that Q' and S' intersect in only two opposite (parallel) sides of them, since in other cases they would coincide. Therefore, in view of (i), S' is a shift of Q' along these opposite sides such that $S' \cap \overline{q'r'} = \{q'\}$ or $\{r'\}$. However, in view of (i) and assuming that Q' and R' are not coplanar, $R' \cap S'$ consists of only one point, either q' or r', which contradicts (ii). Therefore, S' has points in common with Q' only on one side of Q', the same is true for S' and R', considering the symmetry of regular hexagons shown in Fig. 8.6.

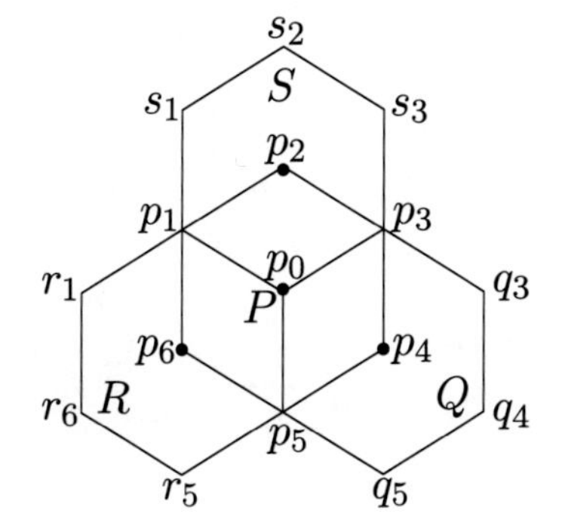

Fig. 8.6 P, Q, R, S are regular hexagons on the same plane, where p_0, p_2, p_4, p_6 are the centers of P, S, Q, R, respectively, and also $Q \cap R \cap S = \{p_0\}$

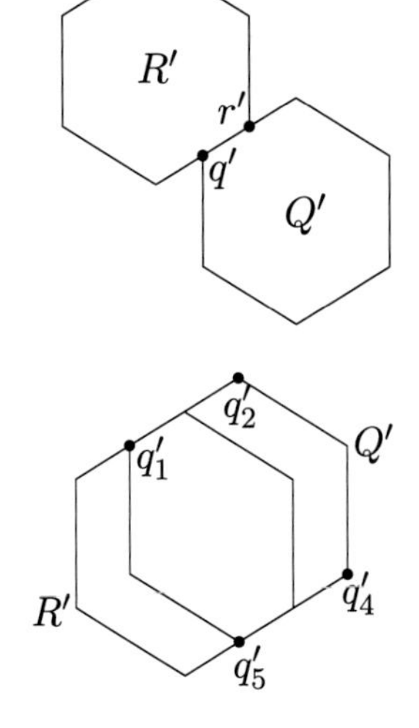

Fig. 8.7 Assuming that the image hexagons Q' and R' are not coplanar, Q' intersects R' only in the nondegenerate line segment $\overline{q'r'}$, where q' is a vertex of Q' and r' is a vertex of R'

Fig. 8.8 This is the case when $Q' \cap R'$ is included in two opposite sides of Q'

Since S' has points in common with Q' on exactly one side of Q' (also for S' and R'), it follows from (i) that $S' \cap \overline{q'r'} = \{q'\}$ or $\{r'\}$. Without loss of generality, assume that $S' \cap \overline{q'r'} = \{q'\}$. In this case, line segment $S' \cap Q'$ is contained in the side of Q' that is adjacent to $\overline{q'r'}$ and has q' as one of its endpoints. This fact implies that q' is the unique intersection of S' with the side of R' containing $\overline{q'r'}$. Since $R' \cap S'$ is an infinite set, we can choose three different and noncollinear points (including q') of $R' \cap S'$. This fact implies that R' and S' are coplanar. However, the plane containing Q' meets the plane containing both R' and S' in two distinct lines, each determined by two sides of Q', where these two sides of Q' meet in q', which contradicts the assumption that Q' and R' are not coplanar. Therefore, Q' and R' should be coplanar, and so should R' and S', given the symmetry of regular hexagons shown in Fig. 8.6.

(d) Now we will prove that $Q' \cap R'$, $R' \cap S'$, and $S' \cap Q'$ are contained in a side of Q', R', and S', respectively. Assume that $Q' \cap R'$ is contained in two opposite sides of Q', say $\overline{q'_1q'_2}$ and $\overline{q'_4q'_5}$. (See Fig. 8.8.)

Because of (i), S' has exactly one point in $Q' \cap R'$, say in $Q' \cap R' \cap \overline{q'_1q'_2}$. Since S' does not coincide with Q', the condition (ii) implies that S' has infinitely many points in $\overline{q'_1q'_2} \setminus Q' \cap R'$ or $\overline{q'_4q'_5} \setminus Q' \cap R'$. (Otherwise, S' contains infinitely many points from some sides of Q' except $\overline{q'_1q'_2}$ and $\overline{q'_4q'_5}$. In this case, S' has infinitely many points in common with Q' on at least two sides of Q', and two of them are not opposite each other. Then it follows from (b) that S' coincides with Q', a contradiction.)

In the first case, i.e., if S' has infinitely many points in $\overline{q'_1q'_2} \setminus Q' \cap R'$, then it follows from (i), (b) and the fact $S' \cap Q' \cap R' \cap \overline{q'_4q'_5} = \emptyset$ that S' should meet Q' in only a part of the side $\overline{q'_1q'_2}$. Therefore, $R' \cap S'$ would be a one-point set, which would contradict (ii).

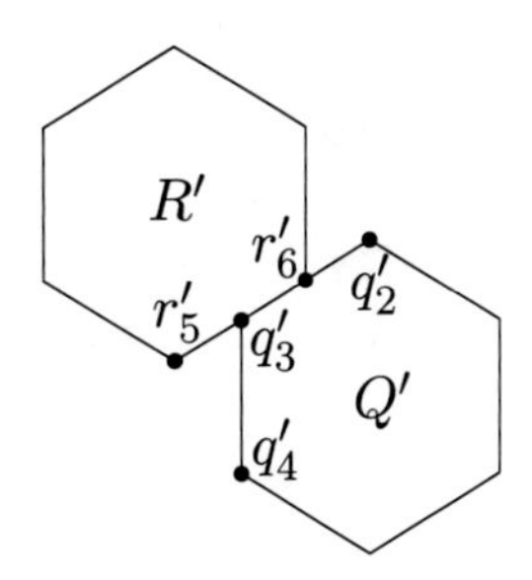

Fig. 8.9 The side $\overline{q'_2 q'_3}$ of Q' partially overlaps the side $\overline{r'_5 r'_6}$ of R', but Q' does not intersect R' anywhere else

In the second case, if $S' \cap \overline{q'_4 q'_5} \setminus Q' \cap R'$ is an infinite set, S' would be a shift of Q' along sides $\overline{q'_1 q'_2}$ and $\overline{q'_4 q'_5}$. (We recall that S' contains points on the side $\overline{q'_1 q'_2}$ of Q'.) Hence, the set $S' \cap Q' \cap R'$ would consist of at least two different points, which would contradict (i). From these arguments and (b), we conclude that $Q' \cap R'$ should be contained in a side of Q'.

By considering the symmetry of Q, R, and S, we can apply the same argument to the case for R' and S' as well as for S' and Q'. Therefore, $R' \cap S'$ should be contained in one side of R', and the same statement should also apply to S' and Q'.

(e) Without loss of generality, we assume that the side $\overline{q'_2 q'_3}$ of Q' partially overlaps the side $\overline{r'_5 r'_6}$ of R' such that the vertex q'_3 of Q' lies on the open side $\overline{r'_5 r'_6}$ of R' and the vertex r'_6 of R' lies on the open side $\overline{q'_2 q'_3}$ of Q', as shown in Fig. 8.9.

According to (d), the infinite set $S' \cap Q'$ must be contained in one side of Q', so by (i), $S' \cap Q'$ is contained either in the line segment $\overline{q'_2 r'_6}$ or in the side $\overline{q'_3 q'_4}$ of Q' that is directly adjacent to $\overline{q'_2 q'_3}$. In both cases, S' can intersect R' in just two points, which contradicts (ii). Therefore, we conclude that $Q' \cap R'$ is the entire side $\overline{q'_2 q'_3} = \overline{r'_5 r'_6}$ of Q' or R'.

Now we can conclude that Q', R', and S' can satisfy the conditions (i) and (ii) if and only if $Q' \cap R' \cap S'$ is the common vertex of these three image hexagons and each of $Q' \cap R'$, $R' \cap S'$ and $S' \cap Q'$ is an entire common side of Q' and R', of R' and S', and of S' and Q', respectively.

These facts imply that Q', R' and S' have the only common vertex $f(p_0)$ and f maps the side $\overline{p_0 p_5}$ of Q onto an adequate side of Q'. Using the same argument and considering the symmetry of regular hexagons shown in Fig. 8.6, we can conclude that f maps each side of Q onto a corresponding side of Q'.

(f) Using an analogous argument presented in the final part of the proof of Theorem 8.39, we conclude that f maps all six vertices of Q onto all six vertices of Q' in the same (or opposite) cyclic order as Q. Similarly, we can verify the same argument for P, R, or S by considering the symmetry of those regular hexagons and the arbitrary choice of Q. We note that $P, Q, R, S \in H_a^m$ and $P', Q', R', S' \in H_b^n$.

(g) Finally, let p_1 and p_2 be arbitrary points of $\mathbb{E}^m$ with $\|p_1 - p_2\| = a$. Then we can select the points p_0, p_3, p_4, p_5, p_6, q_3, q_4, q_5, r_1, r_5, r_6, s_1, s_2 and s_3, as we see in

Fig. 8.6. It then follows from (f) that $\|f(p_1) - f(p_2)\| = b$, which completes the proof. □

In the following corollary, we prove that if a one-to-one mapping $f : \mathbb{E}^n \to \mathbb{E}^n$ maps the periphery of every regular hexagon of side length $a > 0$ onto the periphery of a regular hexagon of side length $b > 0$, then there is an affine isometry $I : \mathbb{E}^n \to \mathbb{E}^n$ such that $f(x) = \frac{b}{a} I(x)$.

Corollary 8.48 *Let n be an integer greater than 1, and let a, b be some positive real constants. If $f : \mathbb{E}^n \to \mathbb{E}^n$ is a one-to-one mapping under which the image of every regular hexagon from H_a^n belongs to H_b^n, then there exists an affine isometry $I : \mathbb{E}^n \to \mathbb{E}^n$ such that*

$$f(x) = \frac{b}{a} I(x)$$

for all $x \in \mathbb{E}^n$.

Proof If we define a mapping $I : \mathbb{E}^n \to \mathbb{E}^n$ by $I(x) = \frac{a}{b} f(x)$ for all $x \in \mathbb{E}^n$, then it follows from Theorem 8.47 that I satisfies $\|I(x) - I(y)\| = a$ for any $x, y \in \mathbb{E}^n$ with $\|x - y\| = a$. By the Beckman-Quarles theorem, I is an affine isometry. □

Mappings Preserving Spheres

Let n be an integer greater than 1. A unit $(n-1)$-sphere is a set of points of $\mathbb{E}^n$, defined as $\{y \in \mathbb{E}^n : \|y - x\| = 1\}$ for some $x \in \mathbb{E}^n$.

The following theorem was proved by S.-M. Jung and B. Kim (see [33]).

Theorem 8.49 (Jung and Kim) *Let n be an integer greater then 1. If a one-to-one mapping $f : \mathbb{E}^n \to \mathbb{E}^n$ maps every unit $(n-1)$-sphere onto a unit $(n-1)$-sphere, then f is an affine isometry.*

Proof We first assume that $n > 2$ and v_1, v_2 are arbitrary points of $\mathbb{E}^n$ with $\|v_1 - v_2\| = 1$. Without loss of generality, we assume that

$$v_1 = \left(\frac{1}{\sqrt{2}}, \frac{1}{\sqrt{2}}, 0, \ldots, 0\right) \quad \text{and} \quad v_2 = \left(0, \frac{1}{\sqrt{2}}, \frac{1}{\sqrt{2}}, 0, \ldots, 0\right).$$

Choose the unit $(n-1)$-spheres $S_1, S_2, \ldots, S_{n+1}$ centered at $a_1 = (\sqrt{2}, 0, \ldots, 0)$, $a_2 = (0, \sqrt{2}, 0, \ldots, 0), \ldots, a_n = (0, 0, \ldots, 0, \sqrt{2})$, and $a_{n+1} = (x, x, \ldots, x)$, respectively, where x is the unique negative real number satisfying $\|a_i - a_{n+1}\| = 2$ for $i \in \{1, 2, \ldots, n\}$. The S_i's are all unit $(n-1)$-spheres such that any pair of these spheres meet each other at exactly one point. Then the same must be true for their image spheres

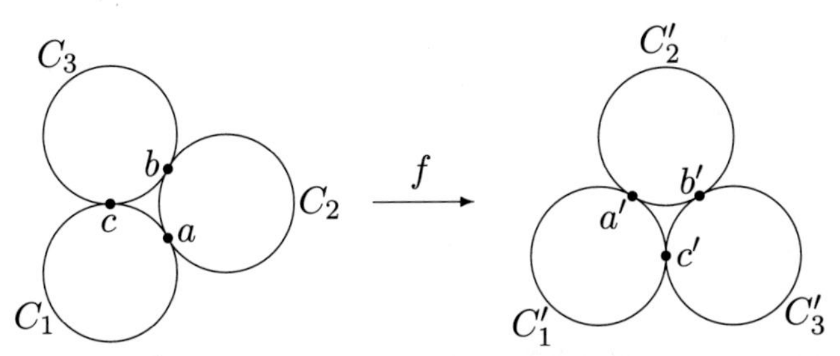

Fig. 8.10 C_1, C_2, C_3 are unit circles such that any two of them intersect in exactly one point. This property also holds for C'_1, C'_2, C'_3, where $C'_i = f(C_i)$

$S'_1, S'_2, \ldots, S'_{n+1}$. We use $a'_1, a'_2, \ldots, a'_{n+1}$ to denote the centers of these image spheres. Since any pair of those image spheres intersect each other at exactly one point, we have $\|a'_i - a'_j\| = 2$ whenever $i \neq j$.

There exists an isometry $\phi : \mathbb{E}^n \to \mathbb{E}^n$ with $\phi(a'_i) = a_i$, and consequently $(\phi \circ f)(S_i) = S_i$ for $i \in \{1, 2, \ldots, n+1\}$. Since $S_1 \cap S_2 = \{v_1\}$ and $S_2 \cap S_3 = \{v_2\}$, we have necessarily $(\phi \circ f)(v_1) = v_1$ and $(\phi \circ f)(v_2) = v_2$. Thus, it holds that

$$\|f(v_1) - f(v_2)\| = \|(\phi \circ f)(v_1) - (\phi \circ f)(v_2)\| = \|v_1 - v_2\| = 1,$$

which implies that if $n > 2$, $f : \mathbb{E}^n \to \mathbb{E}^n$ preserves unit distance.

For $n = 2$, consider two points a and b in $\mathbb{E}^2$ which are separated from each other by the unit distance. We can then draw three unit circles C_1, C_2, and C_3, as shown in Fig. 8.10, so that any two of them touch each other at a point. If we call $C'_i = f(C_i)$ for $i \in \{1, 2, 3\}$, then we get the three contact points a', b', c' which form the three vertices of a unit regular triangle. Since $f(a) = a'$ and $f(b) = b'$, f preserves the unit distance.

Consequently, using the theorem of Beckman and Quarles, we conclude that f is an affine isometry. □

Furthermore, Jung and Kim [32] have proven an interesting theorem regarding the circle-preserving mapping, but it is unfortunate that a detailed proof of this theorem cannot be presented here due to space constraints.

Theorem 8.50 (Jung and Kim) *Let n be an integer greater than 1. If a one-to-one mapping $f : \mathbb{E}^n \to \mathbb{E}^n$ maps every unit circle onto a unit circle, then f is an affine isometry.*

Mappings Preserving Regular Hexahedrons

From now on, a cube means a regular hexahedron whose side length is 1. We first clarify our terms as follows. In Fig. 8.11, we call the points a, b, c, and d *vertices* and the line segments $\overline{ab}$, $\overline{bc}$, $\overline{cd}$, $\overline{da}$ *edges* and the plane bounded by the four edges $\overline{ab}$, $\overline{bc}$, $\overline{cd}$, $\overline{da}$

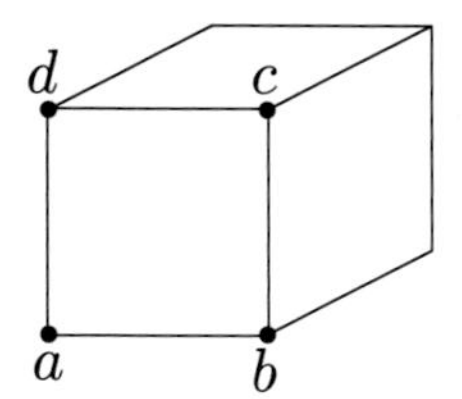

Fig. 8.11 The points a, b, c, d are called vertices, the line segments $\overline{ab}, \overline{bc}, \overline{cd}, \overline{da}$ are called edges, and the plane enclosed by the four edges $\overline{ab}$, $\overline{bc}, \overline{cd}, \overline{da}$ is called a face

face abcd or simply a *face*. Furthermore, by a cube or hexahedron, we mean only the six faces and not the three-dimensional open set bounded by these six faces. Let us denote the three-dimensional open set bounded by the cube A as *inside of* A or simply as Inside(A).

Assume that $p \in A$, where p is a point and A is a cube. First we review the solid angles in three dimensions. If p is a vertex, say $p = a$, then the solid angle subtended by Inside(A) with respect to p is $\frac{1}{2}\pi$. If p is a point that belongs to an edge and is not a vertex, then the solid angle subtended by Inside(A) with respect to p is π. If $p \in A$ is neither a vertex nor an edge point, then the solid angle subtended by Inside(A) with respect to p is 2π. Let us denote the solid angle that Inside(A) subtends with respect to $p \in A$ as $\Omega(A, p)$.

Remark 8.51 Let A be a cube and p a point of A.

(i) $\Omega(A, p) = \frac{1}{2}\pi$ if and only if p is a vertex of A;
(ii) $\Omega(A, p) = \pi$ if and only if p is an edge point of A (and not a vertex);
(iii) $\Omega(A, p) = 2\pi$ if and only if p is neither a vertex nor an edge point of A.

Now we prove the following lemma.

Lemma 8.52 *Let $f : \mathbb{E}^3 \to \mathbb{E}^3$ be a one-to-one mapping that maps every cube onto a cube. It then holds for all cubes A and B: If* Inside$(A) \cap$ Inside$(B) = \emptyset$, *then* Inside$(f(A)) \cap$ Inside$(f(B)) = \emptyset$.

Proof First we check that if $f(b) \in$ Inside$(f(A))$, then $b \in$ Inside(A). In other words, we show that if $b \notin$ Inside(A), then $f(b) \notin$ Inside$(f(A))$. Assume that $b \in A$. Then $f(b) \in f(A)$ and so $f(b) \notin$ Inside$(f(A))$. Assume that $b \notin$ Inside(A) and $b \notin A$. Then we choose another cube B such that $b \in B$ and $B \cap A = \emptyset$. Then $f(B) \cap f(A) = \emptyset$ and therefore $f(b) \notin$ Inside$(f(A))$.

Now let us assume that Inside$(f(A)) \cap$ Inside$(f(B)) \neq \emptyset$. Then, we obtain Inside$(f(A)) \cap f(B) \neq \emptyset$, which implies that $f(b) \in$ Inside$(f(A))$ for some $b \in B$. Therefore, $b \in$ Inside(A) and Inside$(A) \cap B \neq \emptyset$, from which we conclude that Inside$(A) \cap$ Inside$(B) \neq \emptyset$. □

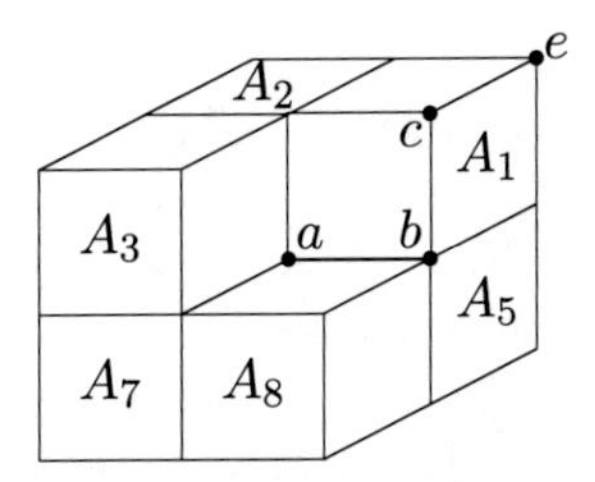

Fig. 8.12 Let a be a vertex of a given cube A_1. Seven more cubes $A_2, A_3, \ldots, A_8$ are then constructed such that a is the common vertex of the eight cubes and $\text{Inside}(A_i) \cap \text{Inside}(A_j) = \emptyset$ for $i \neq j$

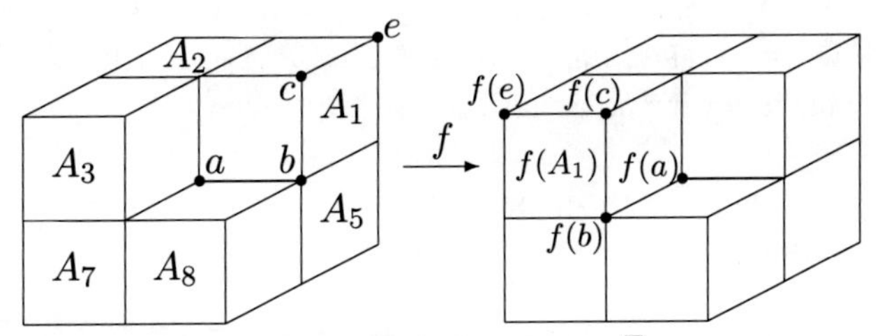

Fig. 8.13 Any two points a and e separated by the distance $\sqrt{3}$ become the two opposite vertices of a cube A_1

Now we prove that it is actually an isometry if a one-to-one mapping preserves regular hexahedrons (see [34]).

Theorem 8.53 (Jung and Kim) *If a one-to-one mapping $f : \mathbb{E}^3 \to \mathbb{E}^3$ maps every cube onto a cube, then f is an affine isometry.*

Proof We show that f preserves the distance $\sqrt{3}$.

Let a be a vertex of a cube A_1. We can then construct seven more cubes A_i, where $i \in \{2, 3, \ldots, 8\}$, such that a is the common vertex of eight cubes A_i, $i \in \{1, 2, \ldots, 8\}$, and $\text{Inside}(A_i) \cap \text{Inside}(A_j) = \emptyset$ for $i \neq j$ (see Fig. 8.12, where A_4 and A_6 are not shown). Then $f(a)$ belongs to $f(A_i)$ for $i \in \{1, 2, \ldots, 8\}$, and by Lemma 8.52, we have $\text{Inside}(f(A_i)) \cap \text{Inside}(f(A_j)) = \emptyset$ for $i \neq j$. Now the solid angle subtended by $\text{Inside}(f(A_i))$ with respect to $f(a)$ is at least $\frac{1}{2}\pi$ for each i, i.e., $\Omega(f(A_i), f(a)) \geq \frac{1}{2}\pi$. Since the maximum solid angle with respect to the point $f(a)$ is 4π, $\Omega(f(A_i), f(a)) = \frac{1}{2}\pi$, and it follows from Remark 8.51 that $f(a)$ is a vertex of $f(A_i)$ for each $i \in \{1, 2, \ldots, 8\}$. Hence, we conclude that if a is a vertex of a cube A, then $f(a)$ is a vertex of a cube $f(A)$.

Any two points a and e separated by the distance $\sqrt{3}$ become the two vertices of the cube A_1, as shown in Fig. 8.13. We then construct seven more cubes $A_2, A_3, \ldots, A_8$ such that the following conditions are met (see Fig. 8.13):

(i) Inside$(A_i) \cap$ Inside$(A_j) = \emptyset$ for $i \neq j$.
(ii) a is the common vertex of the eight cubes $A_1, A_2, \ldots, A_8$.
(iii) Each cube A_i has exactly three vertices (like vertex b), each of which is the common vertex of exactly four cubes. They are all separated by the distance 1 from a.
(iv) Each cube A_i has exactly three vertices (like vertex c), each of which is the common vertex of exactly two cubes. They are all separated by the distance $\sqrt{2}$ from a.
(v) Each cube A_i has exactly one vertex (like vertex e) that belongs to only one cube A_i. It is separated from a by the distance $\sqrt{3}$.

It follows from Lemma 8.52 that Inside$(f(A_i)) \cap$ Inside$(f(A_j)) = \emptyset$ for $i \neq j$. $f(a)$ is the common vertex of the eight cubes $f(A_1), f(A_2), \ldots, f(A_8)$. Thus, as shown on the right side of Fig. 8.13, eight image cubes $f(A_1), f(A_2), \ldots, f(A_8)$ must be placed, so that they have the following properties:

- Each cube $f(A_i)$ has exactly three vertices (like vertex $f(b)$), each of which is the common vertex of exactly four cubes. They are all separated by the distance 1 from $f(a)$;
- Each cube $f(A_i)$ has exactly three vertices (like vertex $f(c)$), each of which is the common vertex of exactly two cubes. They are all separated by the distance $\sqrt{2}$ from $f(a)$;
- Each cube $f(A_i)$ has exactly one vertex (like vertex $f(e)$) that belongs to only one cube $f(A_i)$. It is separated from $f(a)$ by the distance $\sqrt{3}$.

Therefore, we conclude that the distance between $f(a)$ and $f(e)$ is $\sqrt{3}$.

Finally, it follows from the theorem of Beckman and Quarles that f is an affine isometry. □

Remark 8.54 We now conclude this chapter by presenting two interesting open problems:

(i) Does Theorem 8.47 hold for regular heptagons, regular octagons, regular nonagons, etc. instead of regular hexagons?
(ii) Is Theorem 8.53 still valid if $n > 3$? More precisely, it would be interesting to study the two cases where the one-to-one mapping $f : \mathbb{E}^n \to \mathbb{E}^n$ preserves either the 3-dimensional unit cube or the n-dimensional unit cube.

It is interesting to compare Remark 8.54 (i) with the problems presented in paper [21]. The problems presented in [21] were solved in paper [64].

There are also many results published on mappings that preserve unit area, unit perimeter, equilateral triangles, orthogonality of vectors, fixed angles, etc. Related papers include J. Chmieliński [11], J. Lester [45–47], J. Sikorska and T. Szostok [62, 63], A. Koldobsky [43], A. Blanco and A. Turnsek [10], and others.

Correction to: Aleksandrov Problem

Correction to:
Chapter 2 in: S.-M. Jung, *Aleksandrov-Rassias Problems on Distance Preserving Mappings*, Frontiers in Mathematics,
https://doi.org/10.1007/978-3-031-77613-7_2

The original version of the book was inadvertently published with an error in the mathematical symbol on the 5th line from the bottom of page 47. The symbol $\langle \cdot \rangle$ has been replaced with the correct symbol $|| \cdot ||$. The correction had been carried out in Chapter 2.

The updated version of this chapter can be found at
https://doi.org/10.1007/978-3-031-77613-7_2

S.-M. Jung, *Aleksandrov-Rassias Problems on Distance Preserving Mappings*,
Frontiers in Mathematics, https://doi.org/10.1007/978-3-031-77613-7_9

Bibliography

1. A.D. Aleksandrov, Mapping of families of sets. Soviet Math. Dokl. **11**(1), 116–120 (1970)
2. J.A. Baker, Amer. Math. Monthly **78**(6), 655–658 (1971)
3. F.S. Beckman, D.A. Quarles, On isometries of Euclidean spaces. Proc. Amer. Math. Soc. **4**(5), 810–815 (1953)
4. W. Benz, Isometrien in normierten Räumen. Aequationes Math. **29**(1), 204–209 (1985)
5. W. Benz, An elementary proof of the theorem of Beckman and Quarles. Elem. Math. **42**(1), 4–9 (1987)
6. W. Benz, *Geometrische Transformationen* (BI Wissenschaftsverlag, Manheim, 1992)
7. W. Benz, H. Berens, A contribution to a theorem of Ulam and Mazur. Aequationes Math. **34**(1), 61–63 (1987)
8. K. Bezdek, R. Connelly, Two-distance preserving functions from Euclidean space. Period. Math. Hungar. **39**(1–3), 185–200 (1999)
9. R.L. Bishop, Characterizing motions by unit distance invariance. Math. Mag. **46**(3), 148–151 (1973)
10. A. Blanco, A. Turnsek, On maps that preserve orthogonality in normed spaces. Proc. Roy. Soc. Edinburgh Sect. A **136**(4), 709–716 (2006)
11. J. Chmieliński, Stability of angle-preservig mappings on the plane. Math. Inequal. Appl. **8**(3), 497–503 (2005)
12. K. Ciesielski, Th.M. Rassias, On some properties of isometric mappings. Facta Univ. Ser. Math. Inform. **7**, 107–115 (1992)
13. K. Ciesielski, J. Grzybowski, W. Slomczyński, Some remarks on isometries and partitions, in *Problem Book of the Students* (Mathematics of the Jagiellonian University, Kraków, 1980)
14. R. Connelly, Rigidity and energy. Invent. Math. **66**(1), 11–33 (1982)
15. R. Connelly, J. Zaks, The Beckman-Quarles theorem for rational d-spaces, d even and Math, in ed. by A. Bezdek, *Discrete Geometry* (Dekker, New York, 2003), pp.193–199
16. L. Debnath, P. Mikusiński, *Introduction to Hilbert Spaces with Applications*, 2nd edn. (Academic, New York, 2005)
17. B.V. Dekster, Non-isometric distance 1 preserving mapping Math. Arch. Math. **45**(3), 282–283 (1985)
18. B. Farrahi, On distance preserving transformations of Euclidean-like planes over the rational field. Aequationes Math. **14**(3), 473–483 (1976)
19. B. Farrahi, A characterization of isometries of rational Euclidean spaces. J. Geom. **12**(1), 65–68 (1979)
20. P. Fischer, Gy. Muszély, On some new generalizations of the functional equation of Cauchy. Canad. Math. Bull. **10**(2), 197–205 (1967)

S.-M. Jung, *Aleksandrov-Rassias Problems on Distance Preserving Mappings*,
Frontiers in Mathematics, https://doi.org/10.1007/978-3-031-77613-7

21. R.J. Gardner, R.D. Mauldin, Bijections of Math onto itself. Geom. Dedicata **26**, 323–332 (1988)
22. W. Hibi, The Beckman-Quarles theorem for rational spaces: mappings of Math to Math that preserve distance 1. Turkish J. Comput. Math. Ed. **12**(7), 1913–1917 (2021)
23. W. Hibi, The Beckman-Quarles theorem for rational spaces: mappings of Math to Math that preserve distance 1. Turkish J. Comput. Math. Ed. **12**(7), 1918–1922 (2021)
24. W. Hibi, The Beckman-Quarles theorem for rational spaces: mappings of Math to Math that preserve distance 1. Turkish J. Comput. Math. Ed. **12**(7), 1938–1949 (2021)
25. S.-M. Jung, Mappings of conservative distances. Bull. Korean Math. Soc. **40**(1), 9–15 (2003)
26. S.-M. Jung, Mappings preserving some geometrical figures. Acta Math. Hungar. **100**(1–2), 167–175 (2003)
27. S.-M. Jung, Inequalities for distances between points and distance preserving mappings. Nonlinear Anal. **62**(4), 675–681 (2005)
28. S.-M. Jung, On mappings preserving pentagons. Acta Math. Hungar. **110**(3), 261–266 (2006)
29. S.-M. Jung, A characterization of isometries on an open convex set. Bull. Braz. Math. Soc. (N.S.) **37**(3), 351–359 (2006)
30. S.-M. Jung, A characterization of isometries on an open convex set, II. Bull. Braz. Math. Soc. (N.S.) **40**(1), 77–84 (2009)
31. S.-M. Jung, *Hyers-Ulam-Rassias Stability of Functional Equations in Nonlinear Analysis*, Springer Optimization and Its Applications, vol. **48** (Springer, Berlin, 2011)
32. S.-M. Jung, B. Kim, Unit-circle-preserving mappings. Int. J. Math. Math. Sci. **2004**(66), 3577–3586 (2004)
33. S.-M. Jung, B. Kim, Unit-sphere preserving mappings. Glas. Mat. **39**(**59**)(2), 327–330 (2004)
34. S.-M. Jung, B. Kim, Mappings preserving regular hexahedrons, Int. J. Math. Math. Sci. **2005**(21), Article ID 175878, 3511–3515 (2005)
35. S.-M. Jung, K.-S. Lee, Distance-preserving mappings on restricted domains. J. Korea Soc. Math. Ed. (Ser. B) **10**(3), 193–198 (2003)
36. S.-M. Jung, K.-S. Lee, An inequality for distances between $2n$ points and the Aleksandrov-Rassias problem. J. Math. Anal. Appl. **324**(2), 1363–1369 (2006)
37. S.-M. Jung, D. Nam, An inequality for distances among five points and distance preserving mappings. J. Math. Inequal. **12**(4), 1189–1199 (2018)
38. S.-M. Jung, D. Nam, An inequality for distances among n points and distance preserving mappings. J. Math. Inequal. **13**(4), 969–981 (2019)
39. S.-M. Jung, Th.M. Rassias, On distance-preserving mappings. J. Korean Math. Soc. **41**(4), 667–680 (2004)
40. S.-M. Jung, Th.M. Rassias, Mappings preserving two distances. Nonlinear Funct. Anal. Appl. **10**(5), 717–723 (2005)
41. Pl. Kannappan, *Functional Equations and Inequalities with Applications* (Springer, New York, 2009)
42. R.H. Kasriel, *Undergraduate Topology* (W. B. Saunders, Philadelphia, 1971)
43. A. Koldobsky, Operators preserving orthogonality are isometries. Proc. Roy. Soc. Edinburgh Sect. A **123**(5), 835–837 (1993)
44. H. Lenz, Der Satz von Beckmann-Quarles im rationalen Raum. Arch. Math. **49**(2), 106–113 (1987)
45. J.A. Lester, The Beckman-Quarles theorem in Minkowski space for a spacelike square-distance. Arch. Math. **37**(1), 561–568 (1981)
46. J.A. Lester, Euclidean plane point-transformations preserving unit area or unit perimeter. Arch. Math. **45**(6), 561–564 (1985)
47. J. A. Lester, Distance preserving transformations, in ed. by F. Buekenhout, *Handbook of Incidence Geometry* (North Holland, Amsterdam, 1995), pp. 921–944

48. Y. Ma, The Aleksandrov problem for unit distance preserving mapping. Acta Math. Sci. **20**(3), 359–364 (2000)
49. J. Matoušek, *Lectures on Discrete Geometry*. Graduate Texts in Mathematics, vol. **212** (Springer, New York, 2002)
50. S. Mazur, S. Ulam, Sur les transformationes isométriques d'espaces vectoriels normés. C. R. Acad. Sci. Paris **194**, 946–948 (1932)
51. B. Mielnik, Th.M. Rassias, On the Aleksandrov problem of conservative distances. Proc. Amer. Math. Soc. **116**(4), 1115–1118 (1992)
52. B. Nica, The Mazur-Ulam theorem. Expo. Math. **30**(4), 397–398 (2012)
53. F. Rádo, D. Andreason, D. Válcan, Mappings of Math to Math preserving two distances, in *Seminar on Geometry* (Cluj-Napoca,, 1986), pp.9–22
54. Th.M. Rassias, Is a distance one preserving mapping between metric spaces always an isometry? Amer. Math. Monthly **90**(3), 200 (1983)
55. Th.M. Rassias, Query No. 319. Notices Am. Math. Soc. **32**, 9 (1985)
56. Th.M. Rassias, Some remarks on isometric mappings, Facta Univ. Ser. Math. Inform. **2**, 49–52 (1987)
57. Th.M. Rassias, Mappings that preserve unit distance. Indian J. Math. **32**, 275–278 (1990)
58. Th. M. Rassias, Properties of isometries and approximate isometries, in ed. by G.V. Milovanovic, *Recent Progress in Inequalities* (Kluwer, Alphen aan den Rijn, 1998), pp. 341–379
59. Th.M. Rassias, P. Šemrl, On the Mazur-Ulam theorem and the Aleksandrov problem for unit distance preserving mappings. Proc. Amer. Math. Soc. **118**(3), 919–925 (1993)
60. Th.M. Rassias, S. Xiang, On mappings which preserve distances and the Mazur-Ulam theorem. Univ. Beograd. Publ. Elektrotehn. Fak. **11**(4), 1–8 (2000)
61. E.M. Schröder, Eine Ergänzung zum Satz von Beckman and Quarles. Aequationes Math. **19**(1), 89–92 (1979)
62. J. Sikorska, T. Szostok, On mappings preserving equilateral triangles. J. Geom. **80**, 209–218 (2004)
63. J. Sikorska, T. Szostok, On mappings preserving equilateral triangles in normed spaces. J. Geom. **85**, 149–156 (2006)
64. B.A. Slomka, On polygons and injective mappings of the plane, in ed. by M. Ludwig et al., *Asymptotic Geometric Analysis*. Fields Institute Communications, vol. 68 (Springer, New York, 2013), pp. 299–312
65. V. Totik, The Beckman-Quarles theorem via the triangle inequality. Adv. Geom. **21**(4), 541–543 (2021)
66. C.G. Townsend, Congruence-preserving mappings. Math. Mag. **43**(1), 37–39 (1970)
67. A. Tyszka, A discrete form of the Beckman-Quarles theorem. Amer. Math. Monthly **104**(8), 757–761 (1997)
68. A. Tyszka, Discrete versions of the Beckman-Quarles theorem. Aequationes Math. **59**(1–2), 124–133 (2000)
69. A. Tyszka, A discrete form of the Beckman-Quarles theorem for rational eight-space. Aequationes Math. **62**(1–2), 85–93 (2001)
70. J. Väisälä, A proof of the Mazur-Ulam theorem. Amer. Math. Monthly **110**(7), 633–635 (2003)
71. S. Xiang, Aleksandrov problem and mappings which preserve distances, in ed. by Th.M. Rassias, *Functional Equations and Inequalities* (Kluwer, Alphen aan den Rijn, 2000), pp. 297–323
72. S. Xiang, Mappings of conservative distances and the Mazur-Ulam theorem. J. Math. Anal. Appl. **254**(1), 262–274 (2001)
73. S. Xiang, On the Aleksandrov problem and Rassias problem for isometric mappings. Nonlinear Funct. Anal. Appl. **6**(1), 69–77 (2001)

74. J. Zaks, A discrete form of the Beckman-Quarles theorem for rational spaces. J. Geom. **72**(1–2), 199–205 (2001)
75. J. Zaks, The Beckman-Quarles theorem for rational spaces. Discrete Math. **265**(1–3), 311–320 (2003)
76. J. Zaks, On mappings of Math to Math that preserve distances 1 and Math and the Beckman-Quarles theorem. J. Geom. **82**(1–2), 195–203 (2005)

Index

S.-M. Jung, *Aleksandrov-Rassias Problems on Distance Preserving Mappings*, Frontiers in Mathematics, https://doi.org/10.1007/978-3-031-77613-7